for ArcGIS 10

GIS TUTORIAL

Spatial Analysis Workbook

David W. Allen

ESRI PRESS
REDLANDS, CALIFORNIA

Ask for Esri Press titles at your local bookstore or order by calling 800-447-9778, or shop online at www.esri.com/esripress.
Outside the United States, contact your local Esri distributor or shop online at www.eurospanbookstore.com/Esri.
Esri Press titles are distributed to the trade by the following:

In North America:
Ingram Publisher Services
Toll-free telephone: 800-648-3104
Toll-free fax: 800-838-1149
E-mail: customerservice@ingrampublisherservices.com

In the United Kingdom, Europe, Middle East and Africa, Asia, and Australia:
Eurospan Group
3 Henrietta Street
London WC2E 8LU
United Kingdom
Telephone: 44(0) 1767 604972
Fax: 44(0) 1767 601640
E-mail: eurospan@turpin-distribution.com

Contents

Preface

Spatial analysis is the problem-solving aspect of GIS. From a cursory evaluation, the tools seem very basic—buffers, overlays, selections—but when combined in a particular sequence they can reveal things about the data that can't be seen in a spreadsheet or chart. With this book, the reader will not create any new data, but will generate new files based on existing data. That's because analysis isn't about creating new data; it's about making existing data say new things. The key is to know the tools well and design the right sequences to bring the big picture into view. *GIS Tutorial 2: Spatial Analysis Workbook* focuses readers on data presentation and equips them with the skills to build "big picture" maps.

This workbook is a compilation of exercises based on the two-volume set *The ESRI Guide to GIS Analysis* by Andy Mitchell. For best results, it is recommended that you read the corresponding pages in *The ESRI Guide to GIS Analysis* before each tutorial.

Tutorials 1–6 correspond with chapters from *The ESRI Guide to GIS Analysis, Volume 1*. These correlations are easily identified because corresponding chapters have the same titles. Tutorial 1 concentrates on symbology and categorization with the viewer in mind. The material in tutorial 2 deals with mapping quantities and recognizing patterns through classification. In tutorial 3 the exercises and content cover density mapping for value comparison. Tutorial 4 focuses on creating boundaries using visual overlays for performing inside–outside analysis. Tutorial 5 addresses the analysis of distance relationships between features. In tutorial 6 readers learn how to represent data fluctuations over time.

Tutorials 7–9 correspond with content from *The ESRI Guide to GIS Analysis, Volume 2*. The exercises in tutorial 7 deal with displaying geographic distribution to aid analysis. Tutorials 8 and 9 address spatial statistics and introduce a higher level of mathematics for establishing a statistical confidence level for analysis.

Each tutorial contains several elements that build context and engage critical thought in order to reinforce the skills learned. Review sections appear after each exercise that recap the actions taken. Study questions are provided to encourage further analysis. "Your turn" tasks appear within certain tutorials to support independent skill demonstration. The workbook also offers "Other real-world examples" throughout, which provide the reader with real-world scenarios in which the skills covered would be utilized. Additionally, the "Independent projects" section outlines six different scenarios for independent projects that build upon skills acquired in the tutorials.

This workbook is geared toward a more advanced readership than the introductory *GIS Tutorial 1: Basic Workbook* and was written for those who want to learn more about analysis tools in ArcGIS and how to use them. It assumes an existing knowledge of ArcMap and ArcCatalog and requires the use of ArcGIS extensions, as well as third-party tools and scripts, which are included on the accompanying disks.

This book comes with a DVD containing exercise and assignment data and a DVD containing a trial version of ArcGIS Desktop 10. You will need to install the software and data in order to perform the exercises and assignments in this book. (If you have an earlier version of ArcView, ArcEditor, or ArcInfo installed, you will need to uninstall it.) The ArcGIS Desktop 10 DVD provided with this book will work for instructors and basic-level students in exercise labs that previously used an ArcView license of ArcGIS Desktop. Instructions for installing the data and software that come with this book are included in appendix D.

For teacher resources and updates related to this book, go to www.esri.com/esripress.

Acknowledgments

As the program coordinator for the GIS degree offerings at Tarrant County College, I see a lot of what they term "nontraditional" students in my classes—students who have been in the workforce for some time and are returning to college for additional education. Nothing impresses or inspires me more than when someone who decides to add college classes to their already busy schedule comes into my classroom and says "Help me be better." It is to these students I give the most recognition.

I'd like to thank ESRI Press and Andy Mitchell for producing the two volumes that inspired this one. We've used those books in class for many years, and I developed these tutorials to reinforce the topics that are presented there. The great people at ESRI include Judy Hawkins and Shelly Sommer, who helped get the manuscript noticed; Arthur Gelmis and Amy Collins, who carefully edited the manuscript; ESRI Press marketing and distribution staff; Dr. Lauren Scott, who helped review the statistics tutorials; and good friend Clint Brown, who always makes this technology seem so approachable.

Thanks also to Doug Zedler of the City of Fort Worth Fire Department and Gini Connolly of the City of Hurst, Texas, for their help with the reviews. And finally, thanks to the City of Euless administration, which allowed its rich GIS datasets to be used in the making of these tutorials. While the data and process are based in reality, all the scenarios are fictional and should not be associated with the City of Euless.

geographic information systems geographic information systems geographic information systems geographic info
phic information systems geographic information systems geographic information systems geographic info
phic information systems geographic information systems geographic information systems geographic info
phic information systems geographic information systems geographic information systems geographic info
phic information systems geographic information systems geographic information systems geographic info
phic information systems geographic information systems geographic information systems geographic info
phic information systems geographic information systems geographic information systems geographic info
geographic info
geographic info
phic info
geographic info
on systems geographic info
information systems geographic info
phic information systems geographic information systems geographic info
phic information systems geographic information systems geographic info
phic information systems geographic information systems geographic info
phic information systems geographic information systems geographic info
phic information systems geographic information systems geographic info

1

Mapping where things are

The simplest form of analysis is to show features on a map and let the viewer do the analysis in the mind's eye. It falls on the cartographer to use various colors and symbols and to group the data in a logical manner so that the viewer can clearly see the information being highlighted. More complex methods involve categorizing the data and designing symbology for each category.

Tutorial 1-1

Working with categories

The most basic of maps simply show where things are without complicated analysis. These can be very useful and enhanced by symbolizing different categories. By symbolizing categories you can show both location and some characteristics of the features.

Learning objectives

- *Work with unique value categories*
- *Determine a display strategy*
- *Group category displays*
- *Add a legend*
- *Set legend parameters*
- *Use visual analysis to see geographic patterns*

Preparation

- *Read through page 29 in Andy Mitchell's* The ESRI Guide to GIS Analysis, Volume 1 *(ESRI Press 1999).*

Introduction

In the realm of geographic analysis, the first and most simple type is visual analysis—just view it. You can display data on a map with various colors and symbology that will enable the viewer to begin to see geographic patterns. But deciding what aspects of the map features to highlight can take some thought.

It may be as important to map where things are not located as it is to map where they are. Visual analysis allows you to see the groupings of features, as well as areas where features are not grouped. All this takes place in the viewer's mind, since visual analysis doesn't quantify the results. In other words, display the data with good cartographic principles and the viewer will determine what, if any, geographic patterns might exist. Viewers aren't explicitly given an answer, but they can determine one on their own.

You can aid the process by determining the best way to display the data, but that will depend on the audience. If the audience is unfamiliar with the type of data being shown or the area of interest on the map, more reference information about the data may need to be included. You may also want to simplify the way the data is represented or use a subset of the data to make the information more easily understood by a novice audience. Conversely, you may make the data very detailed and specific if the audience is technically savvy and familiar with the data.

In making the map, you will decide what features to display and how to symbolize them. Sometimes simply using the same symbol to show where features are will be enough. For example, seeing where all the stores are in an area might give you an idea of where the shopping district is. All you really want to know is where stores are, and not concern yourself with the type of store.

A more complex method of displaying the data is by using categories, or symbolizing each feature by an attribute value in the data. This also requires a more complex dataset. Your dataset will need a field to store a value describing the feature's type or category. A dataset of store locations may also have a field storing the type of store: clothing, convenience, auto repair, supermarket, fast food, and so on.

For other instances, you may want to see only a subset of the data. These may still be shown with a single symbol but show only one value of a field. It might be fine to see where all crimes are occurring, but the auto-theft task force may only want to see the places where cars have been stolen. The additional data will just confuse the map and possibly obscure a geographic pattern.

When making the map, you can use this "type" or "category" field to assign a different symbol to each value. You might symbolize clothing stores with a picture of a clothes hanger, or a supermarket with a picture of a shopping cart. This still doesn't involve any geographic processing of the data; we're merely showing the feature's location and symbolizing it by a type. The analysis is still taking place in the viewer's mind.

Scenario You are the GIS manager for a city of 60,000 in Texas, and the city planner is asking for a map showing zoning. The target audience is the city council, and each member is very familiar with the zoning categories and the types of projects that may be built in each area. Council members will frequently refer to this map to see in which zoning category a proposed project may fall and what effect that project may have on adjacent property. For example, a proposed concrete mixing plant would only be allowed in an industrial district and would adversely affect residential property if it were allowed to be adjacent.

Since you will have a technically savvy audience, you can use a lot of categories and not worry about the map being too confusing or hard to read. The city planner asks you to use colors that correspond to a standard convention used for zoning. Later, you'll work with setting categories and symbology.

Data The first layer is a zoning dataset containing polygons representing every zoning case ever heard by the city council. An existing field carries a code representing the zoning category assigned to each area. This was already set as the "value field" to define the symbology, so you will only deal with how the categories are shown. This data was created by the city, and a list of the zoning codes and what they mean, called a data dictionary, is provided later in the chapter. You also may find similar datasets from other sources with zoning classifications, and it is important to get their data dictionaries as well.

Start ArcMap and open a map document

1 From the Windows taskbar, click Start > All Programs > ArcGIS > ArcMap. Depending on how ArcGIS and ArcMap have been installed, or which Windows operating system you are using, there may be a slightly different navigation menu from which to open ArcMap. You may also have an ArcMap icon on your desktop, which, when double-clicked, will start ArcMap.

2 In the resulting ArcMap window, click the Existing Maps header and then click "Browse for more...".

3 Browse to the drive on which the tutorial data has been installed (e.g., C:\ESRIPress\GIST2\Maps), click Tutorial 1-1.mxd, and then click Open. This will add the map document to your Existing Maps list. These should be your most commonly used maps.

The data shown is zoning categories, each colored differently. Note that your toolbar locations may differ from those shown. The city planner, for whom you are doing this work, uses this map to visually determine the zoning classification of property.

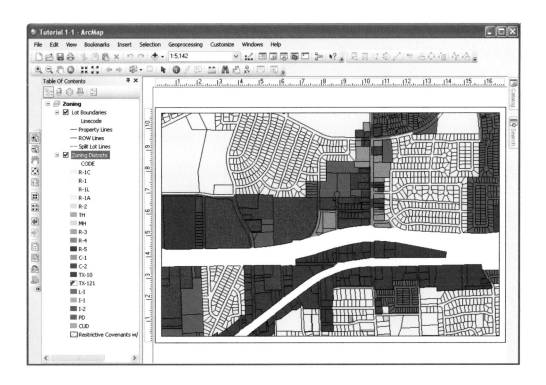

Open a layer attribute table

1 In the table of contents, right-click the Zoning Districts layer and select Open Attribute Table. The data is shown in a tabular format. The columns are the fields in which the data is stored for this layer, and the rows represent each individual feature in the layer.

The field CODE contains the zoning category for each parcel, representing various single-family, multifamily, commercial, and industrial uses. Other fields contain the ordinance number (ORDNO) from the zoning case, a display marker, and the acreage.

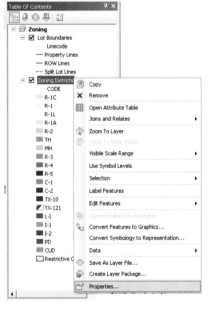

2 Close the attribute table.

Change a category label

The layer came with the code values displayed for each category. While this may be fine for someone who works with these codes every day and recognizes the underlying meaning, someone not familiar with them will want a more descriptive label. These can be changed in the symbol properties of the layer.

1 In the table of contents, right-click the Zoning Districts layer and click Properties.

2 Click the Symbology tab. Note that the Categories selection is set to Unique values and the Value Field is set to CODE. This means that every unique value of the field CODE will be represented in the legend.

To a city planner or someone experienced with zoning districts, these codes would tell the whole story. But for the layperson, they can be difficult to interpret. You will change them to match the simplified list provided by the city planner.

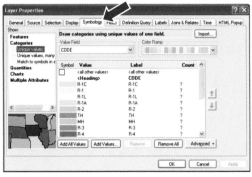

3 Click the R-1C entry under the Label column and type in a new description: **Single Family Custom Dwellings**.

4 Click OK to close the dialog box. The text you typed in the Label column is what will be shown in the legend.

YOUR TURN

Change all the labels to match the list below. **Note:** You can change the width of the column headers Value and Label by dragging the vertical dividers over to reveal the entire category description.

When all the descriptions are entered, click Apply, but do not close the Layer Properties dialog box.

R-1	Single Family Detached Dwellings
R-1L	Single Family Limited Dwellings
R-1A	Single Family Attached Dwellings
R-2	Duplex Dwellings
TH	Townhouses
MH	Mobile Homes
R-3	Multi-Family Dwellings 16 units/ acre
R-4	Multi-Family Dwellings 20 units/ acre

R-5	Multi-Family Dwellings 25 units/ acre
C-1	Neighborhood Business District
C-2	Community Business District
TX-10	Texas Hwy 10 Business District
TX-121	Texas Hwy 121 Business District
L-I	Limited Industrial District
I-1	Light Industrial District
I-2	Heavy Industrial District
PD	Planned Development
CUD	Community Unit Development

All of the descriptions are in, but they are not in the order that the city planner would like. These can be moved in the list to appear in the same order as the list above.

5 Highlight the value R-1A, then click the up arrow to move it above R-1C in the list.

6 Click OK to close the Layer Properties dialog box. Once you have moved the values into the correct order, your table of contents should look like the image to the right.

This zoning layer is used a lot in city maps, so it would be a good idea to save all this work to be used again in the future. This can be done by creating a layer file (layer file names end in ".lyr"). A layer file will save all the symbology settings for a layer, but does not save any of the data. This is useful when a dataset such as zoning needs to be symbolized in several different ways to match different audiences, but you would not want to save the dataset several times. This would lead to a problem in updating the data if you had to make updates in several files and worry about keeping them all current. With a layer file, you maintain one dataset along with several ways that the dataset can be symbolized.

7 Right-click the Zoning Districts layer and select Save As Layer File. Save the file as **DetailedZoning.lyr** in your \GIST2\ MyExercises folder. The next time you want to use the zoning data, don't add the feature class or shapefile; add the layer file and the data will be added already symbolized.

Add a legend to your map layout

Now that you've set everything just right, you can add a legend to the map layout. In fact, even if everything is not exactly how you like, or if the city planner decides to make changes later, any changes you make to the table of contents will automatically be reflected in the legend.

1 On the main menu, click the Insert menu, then click Legend.

2 In the Legend Wizard window, highlight the Lot Boundaries layer in the Legend Items column and click the < button. This will remove the Lot Boundaries layer from the legend.

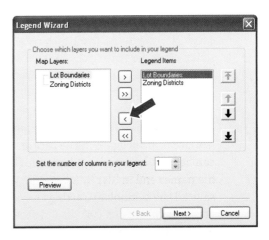

3 Set the number of columns in the legend to 2 and click Next.

4 Change the Legend Title to **City of Oleander**. Click the Color drop-down menu and change the font color to Lapis Lazuli. Click Next.

5 Click the Background color box and select Grey 10% from the drop-down menu. Click Next. Complete the legend with the remaining default values by clicking Next and Finish. The legend is complete and has been added to the map. However, it's a little bit too large and not in the right place. You will use the Select Elements tool ▲ to resize and move it.

When a graphic element is selected, cyan boxes appear at the corners of the element. By clicking these with the Select Elements tool, the element can be altered. Your map contains a specific spot for the legend, so you are going to place it with page coordinates and a set frame size in the legend's properties dialog box.

6 With the Select Elements tool active, right-click the legend and select Properties. Click the Size and Position tab.

7 Enter the new X and Y values for the Position and the new Width and Height values for Size as shown in the graphic below. Click Apply and OK. You may need to click the Zoom Whole Page button on the Layout toolbar to see the entire map. Your completed map should look like the one on the following page. This type of map, showing "where things are," allows viewers to do the recognition and analysis of items in their own mind's eye. You aren't identifying every feature on the map; you are only giving users a color reference they can use to identify features. Likewise, you aren't pointing out any spatial relationships between the data, but letting viewers perform visual analysis to draw their own conclusions.

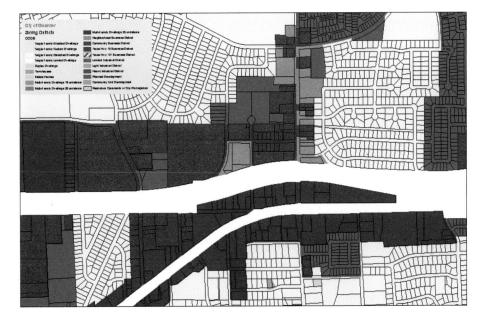

8 Save your map document as **Tutorial 1-1.mxd** in the \GIST2\MyExercises folder. If you are not continuing on to the exercise, exit ArcMap.

Exercise 1-1

The first tutorial showed how to use categories to help the map viewer determine the zoning for a piece of property. The user looks at the property on the map, notes the color, then matches that color in the legend to determine the associated zoning code. You also helped the viewer better understand zoning codes by providing a text translation of the otherwise cryptic values.

In this exercise, you will repeat the process using land-use codes, which show how property is currently being used as opposed to zoning codes, which show the future use of land. These are also somewhat cryptic, so you will change the labels to a simple description. You will also add the legend to the map layout, then resize and move the legend to an appropriate place on the map.

- Continue with the map document you created in this tutorial or open Tutorial 1-1.mxd from the \GIST2\Maps folder.
- Turn off the Zoning Districts layer.
- Add the layer file LandUseCodes.lyr from the \GIST2\Data folder.
- Change the labels for the Land Use values to match the ones at right.
- Make any necessary changes to the legend for it to display all the values.
- Create a layer file for the new symbology.
- Save the results as **Exercise 1-1.mxd** in the \GIST2\MyExercises folder.

A1	Single Family Detached
A2	Mobile Homes
A3	Condominiums
A4	Townhomes
A5	Single Family Detached Limited
AFAC	Airport Facilities
APR	Airport Private Land
AROW	Airport ROW
B1	Multi-Family
B2	Duplex
B3	Triplex
B4	Quadruplex
B5	High Density Multi-Family
CITY	City Property
CITYV	Vacant City Property
CITYW	City Water Utilities Property
CRH	Church
ESMT	Easement
F1	Commercial
F2	Industrial
GOV	Government (State or Federal)
POS	Public Open Space
PRK	Park
PROW	Private Right-of-way
ROW	Right-of-way
SCH	School
UNK	Unclassified
UTIL	Utilities
VAC	Vacant

WHAT TO TURN IN

If you are working in a classroom setting with an instructor, you may be required to submit the maps you created in tutorial 1-1.

Turn in a printed map or screen capture of the following:

Tutorial 1-1.mxd

Exercise 1-1.mxd

Tutorial 1-1 review

You determined that your audience was very familiar with the data being used and decided to make a detailed map of zoning and land use. Next, you checked the attribute table to see which field might contain zoning or land-use data. Since this existed, you were able to use it to set the symbology.

The codes were rather cryptic, even for your expert audience, so you changed the display labels in the symbology tab to have both the code and a text description of each category. This will help remind the viewer what each code represents.

Finally, you added a legend to the map to display the zoning or land-use codes and make the map more complete.

STUDY QUESTIONS

1. Select a piece of property on the map. Can you determine the adjacent zoning category?
2. The city council has been asked to approve a dry-cleaning plant in the Highway 10 business district, but members want to make sure that it is adjacent to other commercial uses and not residential. Can you find a place where this district has residential zoning adjacent to it? How about commercial zoning adjacent to it?

Other real-world examples

Each street in a centerline feature class may have a field with a code representing its status in the thoroughfare plan. You could symbolize the streets with different line thicknesses according to the codes: thick line for freeways, medium line for major collectors, and thin line for residential streets. A quick look at the map would let you determine a fast route based on the amount of traffic a street is designed to handle. You may also do this with speed limit and get a similar result.

A forester may symbolize harvestable tree stands by their age. A darker symbol might be stands ready for harvesting, while a lighter color may be immature trees. A quick view of the map may show where to concentrate logging efforts to minimize the movement of equipment and impact on surrounding areas.

A dataset of all the counties in the United States may have a field representing the political leaning of each county, whether they are Democrats, Republicans, Green Party, Independents, and so on. A candidate for national office may symbolize the map according to these categories to determine friendly areas for fundraising and unfriendly areas for political conversions.

Tutorial 1-2

Controlling which values are displayed

In the previous tutorial, you worked with all the values of a given category field. These were very specific codes that the experts use, but sometimes either for clarity or ease of use in an analysis project, you want to display only some of the categories.

Learning objectives

- *Group similar categories in a legend*
- *Hide categories*
- *Add custom categories*

Preparation

- *Read pages 30–36 in* The ESRI Guide to GIS Analysis, Volume 1.

Introduction

In the previous tutorial, your audience was a technically savvy group that was very familiar with the data. Displaying a long list of zoning or land-use codes was appropriate for them because it allowed them to do the detailed visual analysis required. This isn't always the case, and quite often it may be necessary to simplify the display of the data to fit the audience.

There are two ways to accomplish data simplification. One is to use one of the generalization tools that will physically change, or dissolve, the data to a simpler structure by removing vertices in lines and polygons. This creates a copy of the data in a more simplified form, but also adds to your data-maintenance chores. Any edits of the data will need to be made to both datasets; or perhaps you can delete the simplified data, do your edits, and create the simplified dataset again.

Another method is to control the simplification of the data in the way it is displayed. By grouping values together, you can create the same look as a dissolved dataset without creating a copy of your dataset. This method works well for visual analysis since all analysis takes place in the viewer's mind based on what they see. For more complex analysis types that aim to quantify the results, the dissolve methods are preferred because additional processing may be done to the dissolved dataset that cannot be done to the entire dataset.

Scenario   The maps you made in the first part of this tutorial were a big success with the city council and the city planner. These have caught the eye of the development director, who will be traveling to a developers' conference to promote the city. The director would like to show the zoning map to potential developers, but would like the data simplified to something easier to grasp in a quick overview of the map. Developers will quickly scan the map looking for the type of zoning favorable to their business, and only if they see what they like will they stop to talk details. You need to produce the zoning map again, but this time using general categories.

Data   You are going to use the same datasets as in tutorial 1-1. One change has been made in how the data is symbolized. In the previous tutorial, you used a code that included two special zoning districts that demonstrate an "a la carte" zoning style. This zoning style allows developers to define housing density and other special features of their projects to create a custom zoning category prior to obtaining city council approval. For this tutorial, you will use the "translated zoning," which will generalize the special zoning districts into the closest categorical zoning classification. This is stored in the field CODE2.

Change category labels

1   In ArcMap, open Tutorial 1-2.mxd from the location at which the tutorial data has been installed (e.g., C:\ESRIPress\ GIST2\Maps). The map looks similar to the one you finished in the previous exercise.

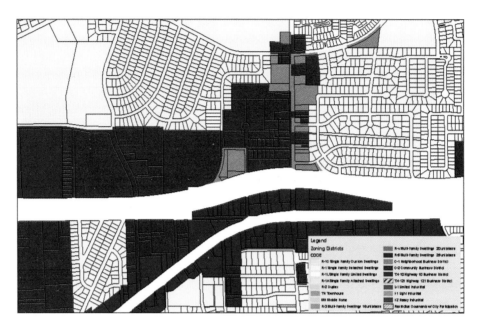

Now you will revise the map for the development director and group all the zoning codes into four categories: Single Family Residential, Multi-Family Residential, Commercial, and Industrial. By doing this, you will make the map suitable for a general audience that can get a quick overview of the city's zoning at a glance without detailed study. Rather than doing this with a dissolve function, you will change the symbology of the layer to match the grouping. It is important to note that you are not changing the underlying data in any way, only changing how it is symbolized.

2 In the table of contents, right-click the Zoning Districts layer and click Properties.

3 Click the Symbology tab. As before, all of the values of the field CODE are displayed and symbolized. You want to create a simple category called "Single Family." This will include these codes: R-1C, R-1, R-1L, R-1A, R-2, TH, and MH.

4 Click R-1C, then hold the Shift key and click MH. All of the single-family zoning categories should be highlighted. Right-click anywhere on the highlighted categories and click Group Values. All of the selected values will now be shown with a single symbol. The label becomes all the labels of the grouped values strung together, which is not desirable. Next, you will change that to a simple description.

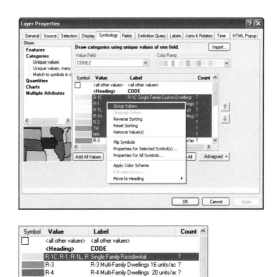

5 Click the label for the grouped values and type **Single Family Residential**.

6 Click Apply and OK to close the dialog box. By default, the color of the first value in the group is used for the combined values. Since single-family residential is typically shown in yellow, you can leave this alone.

YOUR TURN

As you did with the Single Family Residential, create group symbols for the following code values:

   Group R-3, R-4, and R-5. Change the label to Multi-Family Residential.
   Group C-1, C-2, TX-10, and TX-121. Change the label to Commercial.
   Group L-I, I-1, and I-2. Change the label to Industrial.

Click OK to accept the changes and close the dialog box.

When you are done, the table of contents should look like the image to the right.

Next, you will create a layer file of the new symbology for later use.

7 Right-click the Zoning Districts layer and select Save As Layer File. Save it in the \GIST2\MyExercises folder and call it **Grouped Zoning.lyr.** The legend reflects the changes, but looks a little strange. You set the legend to display two columns when there were lots of values to show, but the simplified legend doesn't require two columns.

Modify the legend

1 In the layout, right-click the legend and click Properties.

2 Go to the Items tab, change the number of columns to 1, and then click OK. The legend will display the grouped symbology along with your updated labels in a single column. This should be simple enough for the novice audience to understand.

3 Take a look at the map, which now looks very different from the more detailed version earlier. If you want to zoom into the layout so you can see the legend better, use the Zoom In button on the Layout toolbar. Use the Zoom Whole Page button on the Layout toolbar to return the map to the original view.

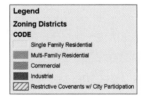

4 Save your map document as **Tutorial 1-2.mxd** in the \GIST2\MyExercises folder. If you are not continuing on to the exercise, exit ArcMap.

Exercise 1-2

The previous exercise demonstrated how to simplify the look of your map by grouping symbol values.

In this exercise, you will repeat the process using land-use codes. These are also somewhat cryptic, so you will change the labels to a simple description. You will also add the legend to the map layout, then resize and move the legend to an appropriate place on the map.

Continue with the map document you created in this tutorial or open Tutorial 1-2.mxd from the \GIST2\Maps folder.

- Turn off the Zoning Districts layer.
- Add the layer file LandUseCodes.lyr from the \GIST2\Data folder.
- Group some of the land-use values together and update the labels to match the definitions below. For values that are not adjacent to each other in the list, select the first one and hold the Control key (Ctrl) down while selecting the others.

A1, A2, A3, A4, A5	Single Family Residential
B1, B2, B3, B4, B5	Multi-Family Residential
F1	Commercial
ESMT, F2, UNK, UTIL	Industrial
POS, PRK	Parks
CITY, CITYV, CITYW, GOV	Government
AROW, PROW, ROW	Right-of-way
AFAC, APR	D/FW International Airport
CRH	Church
SCH	School
VAC	Vacant

- Make any necessary changes to the legend for it to display all the values.
- Create a layer file from the new symbology.
- Save the results as **Exercise 1-2.mxd** in the \GIST2\MyExercises folder.

WHAT TO TURN IN

If you are working in a classroom setting with an instructor, you may be required to submit the maps you created in tutorial 1-2.

Turn in a printed map or screen capture of the following:

        Tutorial 1-2.mxd

        Exercise 1-2.mxd

Tutorial 1-2 review

This time your intended audience was not familiar with all the nuances of zoning and land use, so you simplified the display of the map. You could have simplified the data if more analysis steps were required. However, since you are only doing visual analysis, grouping the symbology will produce the required map without altering the data.

The simplified legend looks cleaner and clearer than the more complex legend, and you had to change how it was displayed. A simple legend with five values was suitable for your audience.

STUDY QUESTIONS

1. Take a look at the map and count how many different zoning categories you can identify. How complex a question could you answer with this level of data?

2. A developer is looking for a commercial corridor in which to locate a business. Can you get a general idea of where the developer should start looking?

3. The city council is reviewing an ordinance to require special fencing between light-industrial and heavy-industrial zones. Is this map suitable for that purpose? Why or why not?

Other real-world examples

Police department data may include up to eight categories of burglaries. While this is great for a detailed study, the map produced for the general public might group all of these into one category.

The Dallas Area Rapid Transit produces detailed maps for each bus route, but when producing a system-wide map, it groups all bus routes into one symbol. Similarly, it groups all train routes into one symbol and all light-rail lines into one symbol. This simplified map reads very well on a regional scale, although it may only answer general questions and not guide you to a specific bus or train.

Tutorial 1-3

Limiting values to display

Legends are the key to how the person creating a map wants the viewer to interpret it. Sometimes all the data is shown in a complex form, and sometimes the data is simplified but still shows the entire dataset. It may also be necessary to display only a subset of the data in order to have the map convey the desired message.

Learning objectives

- *Add only specific values to the legend*
- *Work with legend properties*
- *Work with definition queries*

Introduction

You've seen how to group a number of values together to simplify the way data is displayed on a map, but there may be other times when groups of values do not need to be shown at all. In other words, you may want to keep some of the values from ever showing on the map. Perhaps you want to show only auto-related crime and disregard all other types of crime, or maybe show only major roads and not show local and residential streets. In this tutorial, you will look at ways to restrict which data is symbolized on your maps.

Scenario The city planner has looked at the maps you produced earlier and now has another idea for more maps that you could do. You will display only the single-family zoning categories on the zoning map to produce an inventory of residential zoning. This means that you need to remove all other types of zoning from the map and display only the requested values.

Data You are going to use the same datasets as in tutorial 1-1, but this time the datasets will not have a symbology schema already set. You will need to set the symbology value field to CODE and manually define what values to show.

Set symbol values

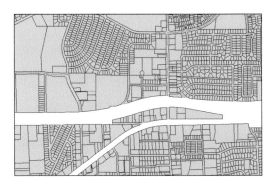

1 In ArcMap, open Tutorial 1-3.mxd from the \GIST2\Maps folder. This time, the map has no symbology set. You'll need to set up controls that will display only the data meeting the request that's been made.

The residential zoning inventory requires that you only display the following values: R-1, R-1A, R-1L, R-2, TH, MH.

These represent the single-family zoning types. The first method you will use will be to add only these values to the symbology editor.

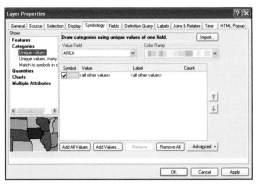

2 In the table of contents, right-click the Zoning Districts layer and click Properties.

3 Click the Symbology tab. In the Show box, click Categories, and then click Unique values.

You need to set the Value Field to the attribute field that contains the zoning classifications. You know from the previous tutorial that it is the CODE field.

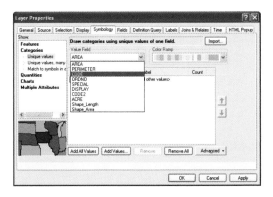

4 Click the Value Field drop-down list and choose CODE.

5 Click the Add Values button. **Note:** The Add All Values button will add each unique occurrence of the Value Field to the legend, while the Add Values button will allow you to select specific field values.

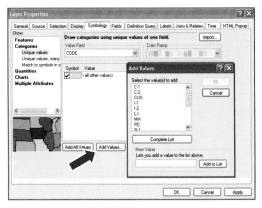

6 From the list, highlight the desired single-family residential codes (listed previously). Hold the Control key to select multiple values. If you don't see a value, click the Complete List button to refresh the list. When you have the right codes selected, click OK. The symbology editor will now show only the selected values. If you missed one, click Add Values again and select it.

7 As in the previous tutorials, change the labels to a simple description of the zoning categories. (Refer to the image for the label names.) Also, clear the <all other values> option to prevent the unselected values from being displayed. Next, you will deal with the color choices.

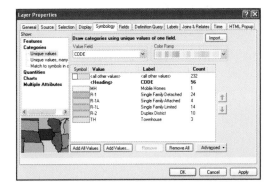

8 Double-click the color symbol to the left of the MH value. This will open the Symbol Selector. Click Yellow, then click OK.

9 Repeat the process and set the other colors as follows:

R-1	Medium Yellow
R-1A	Tan
R-1L	Beige
R-2	Orange
TH	Lt. Orange

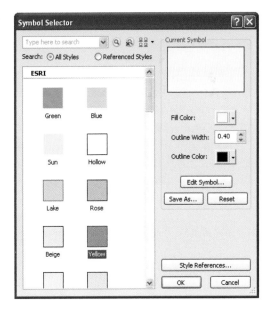

10 Verify that your dialog box matches the graphic, then click OK. The resulting map will show only the specified residential zoning categories in the colors you chose.

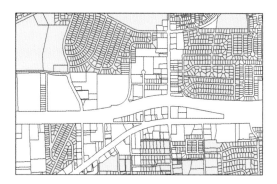

11 Create a layer file of the new symbology. Right-click the Zoning Districts layer, then select Save As Layer File. Save it in the \GIST2\MyExercises folder and call it **ResidentialZoning.lyr.**

YOUR TURN

Add a single column legend to this map displaying the values you selected and symbolized. There are two things that will be different from legends you have worked with in previous tutorials:

1 The legend background will be clear. Open the legend properties, go to the Frame tab, and set the background to 10% Grey.

2 The edge of the legend box will be right along the text and graphics of the legend. To increase the space between them, set the background gap for X and Y to 4 points.

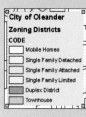

There are many legend parameters that are user-definable. Browse the legend properties box and experiment with a few.

You were able to display only the categories the city planner wanted and learned a little about legends in the process. But there is one problem with this method. The city planner may want you to do some calculations or analysis against the data, and your numbers would reflect the entire dataset, not just the displayed items. Values that are not displayed are still in the database and will still be used in calculations. Take a look at the attribute table to better understand the problem.

Examine the attribute table

1 In the table of contents, right-click the Zoning Districts layer and click Open Attribute Table. The attribute table is displayed and you can see that the CODE column contains values that you did not want shown on the map.

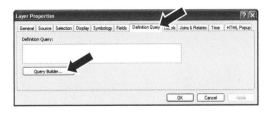

2 Close the Zoning Districts attribute table when you are finished examining it. You need a way to temporarily ignore all values except the single-family residential codes. An easy way to do this is with a definition query. Build a query to select only the values you want and ArcMap will ignore all the other values. It is important to note that the values will not be deleted, only ignored. Removing the definition query will restore all of the dataset values.

Build a definition query

1 In the table of contents, right-click the Zoning Districts layer and click Properties.

2 In the Layer Properties dialog box, click the Definition Query tab and click Query Builder. You will build a query defining the values you wish to see in your map. The format for the query statement will use simple Structured Query Language (SQL) statements composed of "field name, operator, value." You can connect statements together with AND or OR to select multiple values. When you use AND, it means that the statements on both sides of the operator must be true for a feature to be selected. If you use OR, it means that the statements on either side of the operator can be true for the feature to be selected.

Let's say that you wanted only the values of R-1 and R-2. Would the following equation be correct?

[CODE] = 'R-1' AND [CODE] = 'R-2'

In order for a feature to be selected, both of these statements would have to be true. Since a single feature cannot be both R-1 and R-2, no features would be selected. Now try the same statement with OR, where either of the statements can be true for a feature to be selected.

[CODE] = 'R-1' OR [CODE] = 'R-2'

A feature with a value of R-1 would be selected, because at least one of the statements is true, and the same with features containing the value R-2. A feature with a value of I-1 would not be selected because neither of the statements is true.

Your definition query will need to use OR as a connector and select the following values for the field CODE: R-1, R-1A, R-1L, R-2, TH, MH.

The definition query you will build will look like this:

[CODE] = 'R-1' OR [CODE] = 'R-1A' OR [CODE] = 'R-1L' OR [CODE] = 'R-2' OR [CODE] = 'TH' OR [CODE] = 'MH'

3 In the fields box of the Query Builder, double-click [CODE] to add it to the expression box below. Click the equals button (=). Click Get Unique Values and then double-click the value 'R-1'.

4 To string the next selection onto the query, click the "Or" button.

5 Continue to build the query. Double-click [CODE]. Click the equals (=) button and then double-click the value 'R-1A'.

6 Continue until you have built the entire query shown in the graphic. Click the Verify button and let ArcMap validate your query. If there are any errors, go back over the query and check each entry. When the query has been successfully verified, click OK to close the Verify dialog box. Click OK to close the Query Builder and OK again to close the Layer Properties dialog box. Now you will take a look at the attribute table of the Zoning Districts layer and see the results of the definition query.

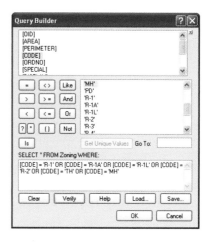

Examine the attribute table again

1 In the table of contents, right-click the Zoning Districts layer and click Open Attribute Table. Notice now that in the CODE column, all the values are among the list you specified. Remember that the definition query did not delete any data. The values not selected by the definition query are only ignored. The process can be reversed simply by removing the definition query.

2 Close the attribute table and save your map document as **Tutorial 1-3.mxd** in the \GIST2\MyExercises folder. If you are not continuing on to the exercise, close ArcMap.

Exercise 1-3

The previous exercise demonstrated how to add only certain field values to your symbology editor, and how to build a definition query to restrict the dataset you are working with.

In this exercise, you will repeat the process using land-use codes. Add only the values necessary and build a definition query to restrict the dataset to the given list.

- Continue with the map document you created in this tutorial or open Tutorial 1-3.mxd from the \GIST2\Maps folder.
- Turn off the Zoning Districts layer.
- Add the dataset LandUse.lyr from the \GIST2\Data folder. Turn on the layer.
- In the symbology editor, set the Value field to UseCode and add only the following values. Change the labels to the correct description.

A1	Single Family Detached
A2	Mobile Homes
A3	Condominiums
A4	Townhomes
A5	Single Family Detached Limited

- Create a layer file of the new symbology.
- Set a definition query to restrict the dataset to only the specified values.
- Add a legend and make any necessary changes to the legend for it to display all the desired values.
- Save the results as **Exercise 1-3.mxd** in the \GIST2\MyExercises folder.

WHAT TO TURN IN

If you are working in a classroom setting with an instructor, you may be required to submit the maps you created in tutorial 1-3.

Turn in a printed map or screen capture of the following:

Tutorial 1-3.mxd

Exercise 1-3.mxd

Tutorial 1-3 review

The city planner gave you the additional task of restricting the dataset to only a specific set of values. You were able to control the display on the map by adding only the values on the list to the symbology editor, and consequently the legend.

A problem with that method is that if you later do any summations or calculations, these will be performed on the entire dataset. One way to solve this is by using a definition query. You built a query to select only those features of interest. Always remember that a definition query does not delete data, but merely causes it to be ignored until the definition query is removed.

A query statement must contain, in this order, a field name, an operator, and then a value. You can string these statements together with AND or OR. Using AND means that **both** sides of the equation must be true to select a feature. Using OR means that **either** side of an equation can be true to select a feature.

STUDY QUESTIONS

1. What additional properties does the legend have?
2. What query would select the zoning values of I-1 and I-2?
3. What query would select the zoning value of R-1 and acreage greater than 50?

Other real-world examples

It is very common to want to work with a subset of your data without having to make a copy.

A city may have within its boundaries many private water wells categorized by which aquifer they tap. You may want to restrict your data to just a single aquifer to calculate the water removal rate from that water source.

The public works department is responsible for repaving the asphalt and concrete streets in the city. You may want to restrict your data to show only the asphalt streets to give them a total linear mileage of streets paved with this material.

You may have a dataset of all fire department responses in a month and want to do some analysis on only the arsons. A definition query on the cause of the fire would accomplish this.

2

Mapping the most and least

GIS data will very often have values associated with it, such as the number of employees at a location or the amount of paper produced at a specific mill. These values may come in many forms, such as a total number of something or the ratio of that value as compared to the whole. The values can be displayed with various symbols or charts that may themselves indicate the impact of the value. Whichever is presented, the data can be mapped to show a relationship.

Tutorial 2-1

Mapping quantities

Mapping the number of things at a given location adds a level of complexity to your map over just mapping where things are. There are several ways to display and interpret quantities in GIS analysis.

Learning objectives

- *Identify discrete features versus continuous phenomena*
- *Work with counts and amounts*
- *Work with ratio normalization*

Preparation

- *Read pages 37–45 in* The ESRI Guide to GIS Analysis, Volume 1.

Introduction

The visual analysis performed in the first tutorial was very helpful in presenting data, but it only showed where certain things were. Another aspect of analysis is to show an amount or quantity associated with the features and let the viewer do a comparison.

There are several ways to work with quantities from the data, including counts, amounts, ratios, and ranks. Knowing how many of something are located at a specific point or the value of the items or their relationship to other locations can add a whole new dimension to comparing features. For example, knowing where businesses are located might show you where the business centers are, but knowing how many workers each business employs would help a transportation planner locate bus stops. The quantity, in this case number of employees, adds additional information to just a simple location map to make it more useful.

When mapping quantities, there are three ways the data may be mapped: as discrete features, as continuous phenomena, or as data summarized by area. Discrete features are those that represent a single item or value. They can be points, lines, or areas, and the quantity associated with them is for that single feature. You may have points that represent store locations with a quantity associated with them of annual sales. Lines could represent street segments with the count of cars that cross it in a day. Areas might represent lakes with the total volume of water they contain. Each is a discrete item with a value for that item.

Continuous phenomena are usually raster datasets and show values continuously over an area. The weather maps that you see on the local news show rainfall over your whole city, even though the rainfall is based on readings from a few rain gauges. You know, however, that rain fell over the entire area, not just at the gauges. Hillshade maps are also continuous phenomena, showing the land slope and elevation over a large area.

The third feature type is data summarized by area. Census data is a familiar type of data that summarizes the number of people who live within an area. Chapter 4 will deal more with data summarized by area and demonstrate how these datasets are created.

Working with quantities is also a good way to start revealing patterns within mapped data. In chapter 1, you merely showed where zoning categories are located. There were no quantities associated with the data that you could use for any meaningful analysis. In this chapter, you will be using quantities to start showing patterns. While mapping the quantities will give a picture from which the viewer may interpret patterns, grouping the quantities into classifications may lead the viewer to a conclusion.

Scenario An investment opportunity has come your way with a grocery store chain to build some specialty stores that target the Hispanic customers in Tarrant County, Texas. These stores have certain brands of food, types of meat, and bakery items that are typically sold in Mexico and are desirable in some U.S. markets. With Texas having once been a part of Mexico, there is still a rich Hispanic heritage in the area, and these markets have become big supporters of the local Hispanic-themed holidays and festivals.

The chain you are interested in is called PePe's Market, which already has five locations in your area. The competition, Buena Vida, has seven locations in the same region. Before you put money into the operation, you want to see if the existing stores are doing well and are located in appropriate markets. You also want to investigate locations for new stores to grow the chain.

You will take some existing data about both chains and match it against some Census 2000 data to see how the stores fare when compared to the ethnicity of neighborhoods. You will be looking for neighborhoods that will seek out stores with specialty goods imported from Mexico as well as customers that will appreciate your community support.

Data The first dataset contains the locations of the existing stores. It's a comma-delimited text file with store names, addresses, number of employees, and average monthly sales compiled from the Web sites of the stores. This will be turned into a point feature class to use in the analysis.

The second dataset is the Census 2000 block-group-level data. This particular set contains a field for the total population, as well as the number of Hispanics in each block group. A study area is already cut out for you to use, but the analysis could be repeated for any area using census data.

The final dataset is the street network data to give the map some reference. It comes from the ESRI Data & Maps media.

Create a layer from x,y coordinates

1 In ArcMap, open Tutorial 2-1.mxd. The layout for your map has been started and a legend added. The map shows Fort Worth, Texas, and the surrounding area. The census block-group-level data has been added, but is not symbolized.

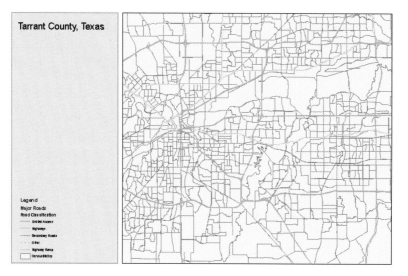

The first thing to do is add the point data representing the store locations. To do this, you will use a tool called "Make XY Event Layer," which can be located with the Search function in ArcMap.

2 On the far right side of your map document, move the cursor over the Search tab, which will open the search dialog automatically.

3 To find the "Make XY Event Layer" tool easily, type the tool's name in the search window and press Enter or click the Search button.

The search dialog box is redesigned in ArcGIS version 10 to give you one single tool to find tools, data, maps, and more. The dialog box can be displayed as a temporary pop-up window by pausing the mouse over the Search tab. Moving the cursor away from the search dialog box will cause it to minimize. Clicking the Search tab will produce a more permanent dialog box, which can be closed by clicking outside the box. The dialog box can be docked to the toolbars by clicking the pushpin icon in the upper right corner of the box. Releasing the pushpin will minimize the dialog box again.

The tutorial used the All setting for the search. The user can also decide to search for specific items by clicking Maps, Data, or Tools. For these to perform efficiently, users should add their data folders to the search index through the Options tab.

Once a search is completed, pausing the cursor over one of the results will display information about that item. This can help identify which map to open or which tool to use.

4 Pause the cursor over the Make XY Event Layer tool to display a short help statement. After reading the help statement, click the tool to open it. Now you will enter the input data, which is a spreadsheet with the stores' longitude and latitude values.

5 Next to the XY Table text box, click the Browse icon and navigate to the exercise data folder and select FoodStoresHispanic.csv. Click Add.

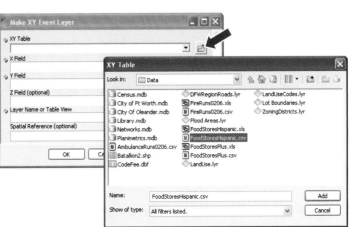

6 Set the X Field and Y Field to the longitude and latitude. Click the drop-down box and select bg_long for X Field and bg_lat for Y Field.

Finally, you will need to set the spatial reference. Longitude and latitude values are not projected coordinates but rather unprojected grid values. In order for ArcMap to project these on the fly to match the project's spatial reference, you need to identify the spatial reference of the input data. You will use an unprojected world coordinate system called WGS 1984. If you are new to projections, see Understanding Projections in the ArcGIS Desktop Help.

7 Click the button next to Spatial Reference, then click Select.

8 Double-click Geographic Coordinate Systems.

9 Double-click World and select WGS 1984, which is a global projection suitable for coordinates in longitude and latitude. Click Add and OK.

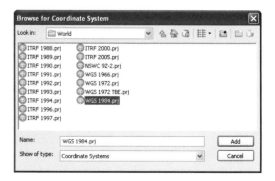

10 When your screen matches the graphic, click OK. You will get a warning that there is a datum conflict between the map and the new layer file. This simply means that the projections of the two datasets are not the same, but ArcMap will handle the datum conflict on the fly.

This process has made a point layer showing the locations of the existing Hispanic-themed grocery stores and has transferred the fields in the table into the new layer as fields. Note that this layer is temporary and will not be saved to the hard disk unless you specifically do so. Now, open the attribute table and see what data has transferred with the points.

Results

☐ 🖳 Current Session
   ☐ ⤬ Make XY Event Layer [183736_02032010]
          ☐ ⦂⦂ Layer Name or Table View: FoodStoresHispanic_Layer
        ⊞ ◇ Inputs
        ⊞ 🗂 Environments
        ☐ ⦂⦂ Messages
              ⓘ Executing: MakeXYEventLayer C:\ESRIPress\GIST2\Data\FoodStoresHispanic.csv bg_long bg_lat FoodStoresHispanic_La
              🕓 Succeeded at Wed Feb 03 18:37:36 2010 (Elapsed Time: 1.00 seconds)

11 Review the Results window and then close it.

Review the attribute table

1 Right-click FoodStoresHispanic_Layer and select Open Attribute Table. The attribute table has the store names and address information, along with the number of employees and the average monthly sales for 2006.

Table

FoodStoresHispanic_Layer

Name	Address	City	State	Zip	NumEmployees	MonthlySales	bg_lat	bg_long
PePe's #322	4200 South Fwy	Fortt Worth	TX	76115	150	156230	32.686362	-97.321162
PePe's #323	245 N. E. 28th Street	Fort Worth	TX	76106	132	129622	32.795309	-97.347191
PePe's #324	1300 E. Pioneer Pkwy	Arlington	TX	76010	161	101250	32.707802	-97.091049
PePe's #325	2700 8th Ave.	Fort Worth	TX	76110	144	148700	32.712844	-97.343962
PePe's #326	3320 Mansfield Hwy	Forest Hill	TX	76119	122	151022	32.681621	-97.277183
Buena Vida #12	4613 E. Lancaster	Fort Worth	TX	76103	143	180560	32.741092	-97.254772
Buena Vida #13	1212 S Ayers	Fort Worth	TX	76105	136	177680	32.73196	-97.271162
Buena Vida #14	3604 Mansfield Hwy	Forest Hill	TX	76119	131	166333	32.678849	-97.270955
Buena Vida #15	1719 8th Ave	Fort Worth	TX	76110	126	171500	32.726258	-97.343536
Buena Vida #16	1300 Lee Ave	Fort Worth	TX	76106	151	133000	32.774693	-97.34857
Buena Vida #17	102 NW 28th St	Fort Worth	TX	76106	153	164590	32.795337	-97.349278
Buena Vida #18	4812 S Freeway	Fort Worth	TX	76115	147	158250	32.67854	-97.3208
Buena Vida #5	3800 Hemphill St	Fort Worth	TX	76110	127	145600	32.692763	-97.331862
Buena Vida #7	2005 N. Riverside Dr	Fort Worth	TX	76111	114	138900	32.788291	-97.301662

⏮ ◀ 1 ▶ ⏭ 🔲 🔳 (0 out of 14 Selected)

FoodStoresHispanic_Layer

2 Close the attribute table.

For the purposes of investigating whether to invest in this company, you will map average sales to see which stores are doing the best. You could put a label next to each point displaying the average sales figure, but that would make it difficult to do a quick visual comparison between stores. Instead, you should use different-sized dots to represent sales. The larger the dot, the higher the sales; the smaller the dot, the lower the sales. A quick scan of the map looking for the largest dots would give the answer. This is another good example of how informative visual analysis can be.

Set a graduated symbols classification scheme

1 Open the properties of FoodStoresHispanic_Layer by right-clicking the layer and selecting Properties.

2 Go to the Symbology tab and select Quantities > Graduated symbols in the Show window.

3 Click the drop-down menu next to the Value field box and select MonthlySales. The symbols will represent the average monthly sales for the past 12 months. The default classification is five classes with a preset size range. In order for these to display better, set the sizes a little larger.

4 Change the Symbol Size values from 4 and 18 to 8 and 25. Try setting the size, then clicking Apply to preview the changes in the map. You may want to experiment with different size and color combinations to get a setup that appeals to you.

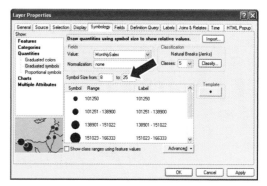

5 When you are satisfied with the results, click OK. The resulting map shows the store locations symbolized by dots that are sized based on the average monthly sales at each store. This is an example of mapping and symbolizing an amount. You can see that some stores have higher sales than others with a quick glance at the map. You can also do some additional visual analysis to see how stores relate to the freeway network.

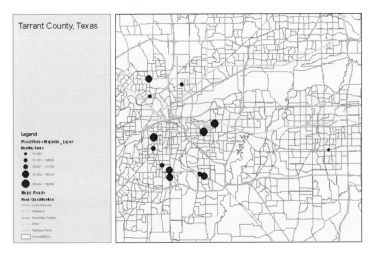

When you sit down and talk with the owners to discuss investing money with them, it would be revealing to quiz them on their expansion plans. This map shows the stores to be grouped in certain areas, and you could explore other areas to see if they would be suitable for additional stores. You know that the stores cater to the Hispanic population, having specialty ethnic foods, so if you can find the areas of the county with a high concentration of Hispanics, this might suggest areas for expansion.

Included in this map is the Census 2000 data for the county. You will take a look at the attribute table and see if there is a field that could be used to find concentrations of Hispanics.

Examine the Census layer attribute table

1 Right-click the CensusBlkGrp layer and select Open Attribute Table.

2 Move the slider bar to the right and you will see a field called Hispanic. This represents the number of Hispanics counted in each census tract, or data summarized by area. You can symbolize each polygon according to the count in the Hispanic field.

Summarizing by area is a common way to show data. It's not possible to locate every person and place a point on the map where they live—the data would be very complex and pretty revealing about individuals. It is more convenient, and a better safeguard of privacy, to set small study areas and summarize the data for that area. In the case of census data, you can see how many people live on a certain block, but you are not able to gain personal information about an individual household.

3 Close the attribute table.

Set the classification scheme

1 Right-click the CensusBlkGrp layer and select Properties. Go to the Symbology tab and select Quantities > Graduated colors in the Show window. **Note:** The graphics show the default color ramp; your colors may be different because ArcMap remembers the last color ramp chosen.

2 In the Value drop-down box, select HISPANIC.

You will be using the count of Hispanics in each census tract to assign a color. The tracts with the darkest color will have the largest Hispanic population counts, and the lighter colors will represent the lower counts.

3 Click Apply to preview the results. Move the dialog box so you can see the map, if necessary. You can try different color ramps, and when you are satisfied click OK. You accepted the default settings for the classification. The default classification method is called natural breaks and displays five categories. The tutorial will show more about this later, but for now it is sufficient to recognize that the darker color represents a larger count, and the lighter color represents a lower count.

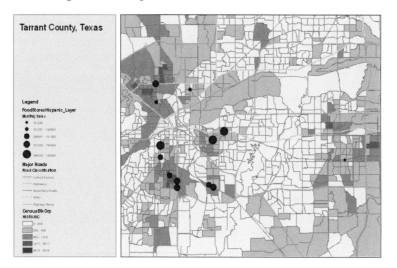

Using visual analysis, you can see that the stores seem to be grouped near the polygons with the darker color. Stop for a moment and think about what you are seeing. The darker areas have a large number of Hispanics, but does that represent a large percentage of the population in that census tract? A tract may score high on this map with a count of 2,500 Hispanics, but what if there were 15,000 people in that census tract? That's only 16.7 percent. At the start of the tutorial it said to look at concentrations, or areas where there are more Hispanics as a percentage of the total population. You could calculate that by dividing the Hispanic count in each tract by the total population of each tract. This is called normalization. We're looking for the population of Hispanics normalized by the total population. For these purposes, you can substitute the words "divided by" for "normalize by." So to normalize by total population means to divide by total population. This is an example of mapping and symbolizing a ratio. You can achieve this by modifying the classification scheme.

Set a normalization field

1 Open the properties of the CensusBlkGrp layer and go to the Symbology tab. Click the Normalization drop-down box and select POP2000.

2 Click OK to accept the changes. Wow, what a difference! Now you are looking at areas that are predominantly Hispanic and would be likely to shop at your specialty grocery store and benefit from the community support the store provides. There are also some areas where the stores are doing well, and there seems to be room for more stores. It also gives some insight into why the easternmost store isn't doing so well. While the count of Hispanics seemed high, it is not in an area of high concentration, which is to say that a large area with a large number of people is not as desirable as a small area with a large number of people. The ratio of potential customers was lower. It is important to be aware of what you are asking for in your analysis. This demonstrated that showing count versus concentration can produce two very different results.

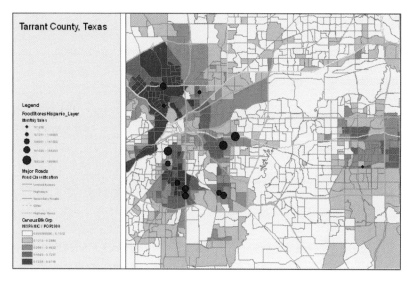

3 Save your map document as **Tutorial 2-1.mxd** in the \GIST2\MyExercises folder. If you are not continuing on to the exercise, exit ArcMap.

Exercise 2-1

The tutorial showed how to create an XY event layer from a comma-delimited text file that contains longitude and latitude values. It went on to demonstrate how to symbolize a dataset with graduated symbols and colors.

2-1
2-2
2-3
2-4

In this exercise, you will repeat the process using different datasets. You are also looking at investing in a different type of specialty grocery store. This one caters to high-end clients that are looking for more exotic foods from throughout the world. Each has been given a product selection rank, representing how many exotic food selections it carries. You will also display store locations over income-level data taken from Census 2000.

- Continue using the map document you created in this tutorial or open Tutorial 2-1.mxd from the \GIST2\Maps folder.
- Turn off the FoodStoresHispanic and CensusBlkGrp layers.
- Make an XY event layer from the file FoodStoresPlus.csv in the data folder. The data is also in longitude/latitude coordinates and will need a spatial reference that will accommodate them.
- Symbolize the FoodStoresPlus point layer with graduated symbols using the ProductSelection field.
- Adjust the map scale so that all the new points in the FoodStoresPlus layer are visible.
- Add the CensusBlkGrpIncome layer to your map document (found in \Census.mdb\ DFWRegion).
- Symbolize the CensusBlkGrpIncome layer with graduated colors using the P053001 field (determined by the Census 2000 metadata) to represent median household income.
- Change the titles, colors, legend, and so on, to make a visually pleasing map.
- Save the results as **Exercise 2-1.mxd** in the \GIST2\MyExercises folder.

WHAT TO TURN IN

If you are working in a classroom setting with an instructor, you may be required to submit the maps you created in tutorial 2-1.

Turn in a printed map or screen capture of the following:

Tutorial 2-1.mxd

Exercise 2-1.mxd

Tutorial 2-1 review

Showing quantities on your maps gives another level of complexity over just showing where things are. You can visualize counts, amounts, and ratios with other locational data shown as an overlay.

The graduated symbols classification scheme is a great way to symbolize quantities associated with point data.

Remember to let the dot sizes tell the story, and not confuse things by trying to make the dots different colors. The viewers will get a much better understanding of the data if they realize all the purple dots are representing different values of the same item. A map full of different-colored dots that vary in size is just confusing and wouldn't read well to viewers.

It is also important to remember what the quantities represent and how to interpret them. As you saw with the normalization feature in the layer properties, you can create ratios on the fly. You used the total number of Hispanics and divided (or normalized) by the total number of people of all ethnic groups. Always make sure when setting a normalization field that it matches a real mathematical equation you would perform. The equation mimicked here was to get the percent of the total (part / total $\times$ 100, or Hispanic population divided by total population).

There is another normalization selection built into the layer properties called Percent of Total. This would divide the number of Hispanics in a particular tract by the total Hispanic population, or the sum of the value field. Try changing that on your map and notice the difference in the results. That's certainly fine to do, but make sure that you understand what the map is showing (the percentage of the total Hispanic population that resides in each tract) before you use it for any purpose.

There is more discussion on classification methods later in this book, but it's interesting to note that these maps read rather well with the default settings and could have been labeled with very basic text such as low/medium/high. The actual numbers were not as important as the relationship between the numbers.

STUDY QUESTIONS

1. Read in the ArcGIS Desktop Help about the difference between graduated symbols and proportional symbols. When would you use one over the other?

2. Would changing the number of classes or the display colors cause the map to show a different answer?

3. What are the responsibilities of the person creating a map to display the data without manipulation?

Other real-world examples

A police department commonly maps locations of auto accidents and uses graduated symbols to display multiple accidents at a single location. This highlights problem intersections.

The Census 2000 data is a treasure chest of quantity data. For a complete understanding of all that census data can provide, see *Unlocking the Census with GIS* by Alan Peters and Heather MacDonald (ESRI Press 2004).

Traffic counts are routinely done for stretches of roads, then associated with the GIS data. The streets are mapped with graduated symbols: the larger the symbol, the higher the count; the thinner the street, the smaller the count.

Tutorial 2-2

Choosing classes

There are many different classification methods to choose from, and the right choice can make or break your map. The data-distribution diagram, sometimes known as the histogram or frequency-distribution chart, will help in making the right choice.

Learning objectives

- *Understand classification schemes*
- *Understand data statistics*
- *Understand data distribution characteristics*

Preparation

- *Read pages 46–55 in* The ESRI Guide to GIS Analysis, Volume 1.

Introduction

When mapping quantities, the data is commonly divided into classes, or ranges of values, to be shown on the map. The methods for putting data into classes, or classification, include natural breaks, quantile, equal interval, and standard deviation. A combination of the desired look of the map and the distribution of the data is used to determine which method to use.

The most common, and the default choice in ArcMap, is the natural breaks method, using the Jenks algorithm. Mathematician George Jenks developed the method of finding natural groupings of data and setting classifications based on those groupings. This groups similar values and maximizes the differences between classes. For most studies, the Jenks natural breaks method is the preferred choice.

The quantile classification method deals directly with the number of features in each class. The number of features is divided by the number of classes you specify, and the resulting quantity is placed in each class. The quantity of each class is the same. This classification works well if you want to show just a certain percentage of the results, for instance the top 20 percent of the values. It also works well if the data is evenly distributed across the entire value range. It is not desirable for data where the values tend to group together or the area of the features is vastly different. Values that are not similar may be placed in the same grouping.

2-1
2-2
2-3
2-4

The equal-interval classification deals with the size of the classes. The total range of the data values is divided by the number of classes you specify. Values are then placed in the classes with no regard to how many values fall within each range. This works well if the range of values is familiar to the audience, such as percentages, but in some instances may place all the values in one or two classes with other classes being empty.

The standard deviation classification will show how much a value varies from the mean. The standard deviation is calculated for each value, and the result is classified as above or below the standard deviation. The data needs to be tightly grouped in a classic bell curve for this classification to be effective. Otherwise you get a significant number of features four and five standard deviations off the norm, which won't show any groupings. The drawback to this classification is that the actual values are not shown, just their relationship to the mean. Very high or very low values can affect the calculation and skew the results.

Step one in deciding which classification to use is to view the data-distribution diagram for the dataset. Evenly distributed data might be a candidate for a quantile or equal-interval classification. Data that is grouped tightly or for which you want to show the median values might be best shown with a standard deviation classification. Data that has one or more distinct groupings is best shown with the Jenks natural breaks classification.

After the classification method is chosen, the number of classes must be determined. The larger the number of classes, the less change there will be between the class values, thus making changes more subtle and harder to detect. This may work well if a more complex map is needed for a sophisticated audience.

Conversely, fewer classes will greatly simplify the data display. A quick glance analysis for a very general audience may call for a simpler map. Sometimes a simple "low, medium, high" split is sufficient.

As you gain more experience with different datasets, classification methods, and viewing audiences, you will begin to develop a feel for which combination will work best.

Scenario   The city planner has given you several datasets and asked you to determine the best classification method for each. Show the city planner what can be done, listing the classification method and why it will or won't work for this analysis.

Data   The first dataset is the census block group data for Tarrant County, Texas. The fields Median_Age and Pop_2004 contain the counts for each census tract.

The second dataset is the parcel data for the City of Oleander, Texas. You will display the values in the Year_Built field, setting up a classification to show the age of housing.

Examine the attribute table for the census layer

1 In ArcMap, open Tutorial 2-2.mxd. You get the familiar map of Tarrant County containing the census data. You will look at the data fields one by one and try to determine the best classifications to use.

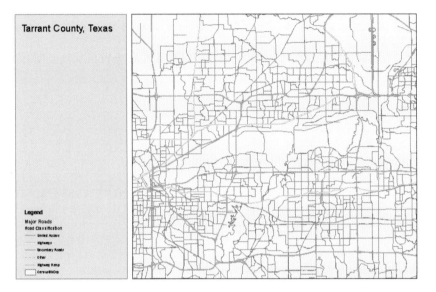

2 Right-click the CensusBlkGrp layer and open the attribute table.

3 Right-click the Population 2004 field and select Statistics. The statistics box will open, showing various statistics about the data contained in this field. The interesting part is the Frequency Distribution chart on the right.

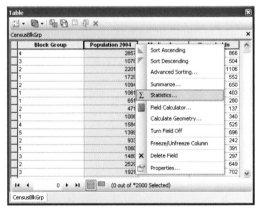

The numbers across the bottom represent the range of values in this field, and the numbers up the side are the number of features with the same value. What you want to do is visualize a curve over the chart display to see how well the data is distributed across the range of values. You see that the data is a regular curve, weighted to the lower end of the scale. As the first test, you will classify the data using the Jenks natural breaks method and let it find the natural groupings of data.

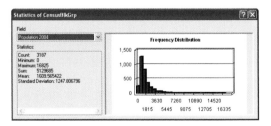

4 Close the Statistics dialog box and the attribute table.

Test the classification scheme for total population

For a support map in a series of analyses showing population characteristics, you need to display the population total in a manner that is easy to interpret. The map should give the viewer an idea of where the largest population centers are, where the smallest population centers are, and provide a range of values in between. If you showed only areas of high or low population, there would not be a good frame of comparison between the values.

1 Right-click the CensusBlkGrp layer and open the properties. Go to the Symbology tab and select Quantities > Graduated colors in the Show window. Set the Value field to Population 2004.

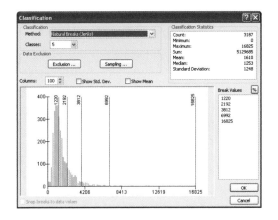

The default classification of Jenks with five classes will be selected. In many instances, you could just use the Classes drop-down menu to adjust the number of classes and continue using the natural breaks classification. But the city planner wants an explanation of the choice, so look into it further.

2 Click the Classify button to open the Classification dialog box. The data values are grouped to the left of the chart, and you can see that the gaps between the classification break lines in blue are smallest at the left and get larger as the values increase. When you view this on the map, you would expect to see a large number of polygons representing the first two classes and just a few representing the rest of the data.

3 Click OK. Change the color ramp if you wish, accept the default classification settings, and click OK to close the Properties dialog box. Examine the results on the map.

Tarrant County, Texas

Legend
Major Roads
Road Classification
—— United Access
—— Highways
—— Secondary Roads
—— Other
—— Highway Ramps
CensusBlkGrp
Population 2004
☐ 0 - 1220
☐ 1221 - 2192
☐ 2193 - 3812
■ 3813 - 6562
■ 6563 - 16828

It's easy to see the groupings of data that correspond to the classifications you set. Some circles have been drawn in the image above for emphasis, but these will not appear on your map. All ranges are represented on the map, and the data is easy to interpret. So you can tell the city planner that the Jenks natural breaks classification will work well for population data.

4 Print or save a screen capture of this map.

Next you will look at the Median Age field and choose a classification for it.

Test the classification scheme for median age

As with the total population map, the map of median ages should highlight both the high and low areas, as well as the medium-range values for comparison.

1 Right-click the CensusBlkGrp layer and select Open Attribute Table. Right-click the Median Age field and select Statistics to open the Statistics dialog box. Draw an imaginary curve along the top of the chart, and you will get the classic bell curve with the data centrally weighted. Try a quantile and an equal-interval classification on this field.

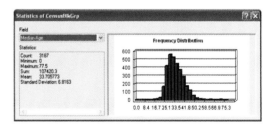

Statistics of CensusBlkGrp

Field
Median Age

Statistics:
Count: 3187
Minimum: 0
Maximum: 77.5
Sum: 107420.3
Mean: 33.705773
Standard Deviation: 6.8163

Frequency Distribution

600
500
400
300
200
100
0
0.0 8.4 16.7 25.1 33.5 41.8 50.2 58.6 66.9 75.3

2 Close the Statistics dialog box and the attribute table.

3 Right-click the CensusBlkGrp layer and open the properties. Change the Value field to Median Age.

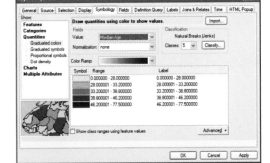

4 Click the Classify button and change the classification method to Equal Interval. As the name implies, the intervals between the classification break lines are equal. You can verify this by looking in the Break Values display. The values are approximately 15 units apart. Next, look at how many features are in each group.

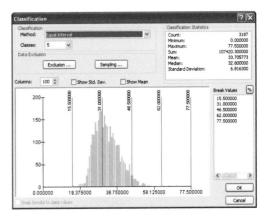

5 Click the first value in the Break Values box. At the bottom of the dialog box, you can see that there are two features in the first class.

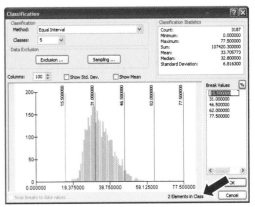

6 One by one, click the break values and look at the resulting number of features in each class. Click OK and OK to close the dialog boxes and see the results of this classification in the map. You will find that the second and the third classes contain most of the features. When displayed on the map, you see only a few of the lowest classifications shown. The majority of the map is represented by only two of the colors. The data doesn't lie; these values are true. But this classification doesn't show any distinct groupings. This classification isn't suitable, so you should try another one.

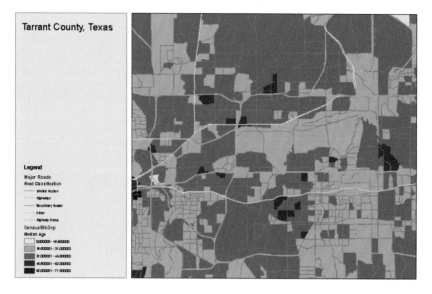

7 Open the layer properties for CensusBlkGroup again and click the Classify button. Change the classification method to Quantile. By looking at the Break Values field, you can see that the first class range is rather large. The next few class ranges are small, then the last class range is again rather large.

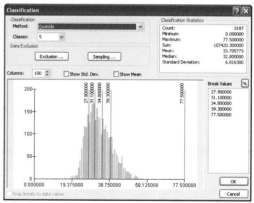

8 Click each of the break values and note how many features are in each class. Quantile means that each class will have close to the same number of features, which can be seen by examining the break value numbers. There are around 645 features in each group.

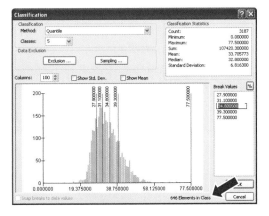

9 Click OK, then OK to close the Properties dialog box and view how the classification setting changes the map display. Some pretty good groupings are apparent with this classification, and all the value ranges are represented. This is appropriate for this dataset.

10 Print or save a screen capture of this map.

Now you will look at the age of houses in Oleander. Each parcel has a year of construction for the house, if there is one. By looking at the data, you might be able to see groupings of years where there might have been a construction surge. The question is, which classification scheme would be the best to display these groupings?

Analyze Year Built data

The data showing in what year each house was constructed can help pinpoint areas in need of renewal. The data needs to be shown in a manner that can be easily understood, and matched to conventions for date-based data.

1 On the main menu, click Bookmarks and select the City of Oleander bookmark. Change the map title to **City of Oleander**. The data is a parcel map of Oleander. Right now the parcels are all the same color, but you will change the classification scheme to show the year of construction for each house. Before you decide on which classifications to consider, you will need to look at the frequency distribution diagram.

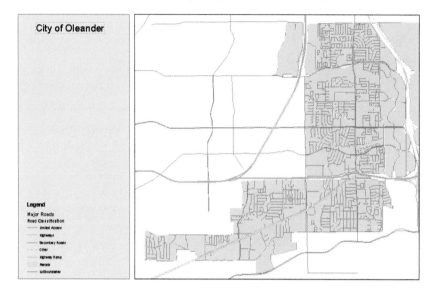

2 Right-click the Parcels layer and open the attribute table. Right-click the field YearBuilt and select Statistics. This looks really odd. There are some values at zero; the rest are way over in the upper ranges. What you are seeing are the values from the parcels for which there is either no structure or the year of construction is not known. These values, called outliers, are skewing the data to the low end of the scale.

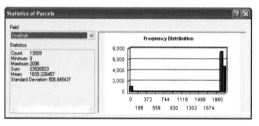

3 Close the Statistics window and the attribute table.

If you did a definition query to remove all the outliers, those parcels would not be included in the frequency distribution diagram and the data would look better. But they also wouldn't display in the map, leaving many gaps. Instead, there's a feature in the properties that instructs ArcMap to ignore them in the classification scheme.

Note: This is a large dataset, and ArcMap will likely reach the maximum number of samples used to set the symbology groupings. The default sample of 500 features will work fine for this dataset, and you can ignore the warnings. If a larger sample size is needed, the value can be changed by clicking the Sampling button on the Classification dialog box.

4 Right-click the Parcels layer and open the properties. Set the Show window to Graduated Colors, change the Value field to YearBuilt, and click the Classify button.

You will use the Data Exclusion function to remove all the values less than 1,900. Then you will review the frequency distribution diagram again and make a decision on which classification method to use.

5 Click the Exclusion button. Double-click the field [YearBuilt], click the < symbol, then type in a space and the number **1900**. When your dialog box matches the image, click OK.

Now look at the difference in the frequency distribution diagram. The features that meet the Data Exclusion query are no longer considered in the frequency distribution diagram or used in determining the classes and values for the classification.

The data spans from 1926 to 2006, or 80 years. The natural breaks method isn't really classifying the data in any manner that works well with dates. The ranges span anywhere from 6 to 15 years, making comparisons difficult. What if you set

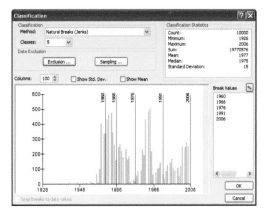

the intervals to 10 years to make them easier to understand? To do that, you would set the classification method to Equal Interval and the number of classes to 8. Dividing the 80-year span into 8 equal intervals would make the intervals 10 years each.

6 Change the classification method to Equal Interval and set the number of classes to 8. Examine the frequency distribution diagram. Click OK.

Change the symbol outline

1 Right-click one of the colored boxes in the Symbol column and select Properties for All Symbols.

2 In the Symbol Selector, set the Outline Width to 0. Click OK, then OK again to close the properties. The result shows clearly the progression of construction across Oleander over the past 80 years in time ranges that make sense to a broad audience.

But the map can be made better. While it makes sense that the class intervals are 10 years, it would make better sense if the intervals started and stopped on the decades. The automated classification methods wouldn't do this, but you can do it manually.

Set a manual classification

1 Right-click the Parcels layer and open the properties. Click the Classify button. Change the number of classes to 9.

2 In the Break Values box, type in new values representing the decades, starting with 1929 and ending with 2009.

3 Click OK, then OK again to close the dialog box. Now the ranges in the legend represent whole decades, making the map fit with a standard convention of how people think of years. This will be easily understood by everyone.

4 Print or make a screen capture of this map. Save your map document as **Tutorial 2-2.mxd** in the \GIST2\MyExercises folder. If you are not continuing on to the exercise, exit ArcMap.

Exercise 2-2

The tutorial showed various classification methods and why they were chosen for particular datasets.

In this exercise, examine the field DU in the Parcels layer. This represents dwelling units. Exclude the zero values, which are vacant lots, and select a classification method that will give the viewer an idea of areas with one or two dwelling units per parcel, areas with a mid range of dwelling units per parcel, and areas with 50 or more dwelling units per acre.

- Continue with the map document you created in this tutorial or open Tutorial 2-2.mxd from the \GIST2\Maps folder.
- Go to the City of Oleander bookmark.
- Examine the Frequency Distribution Diagram for the DU field in the Parcels layer.
- Change the symbology scheme to graduated colors, if necessary.
- Change the Value field, the classification method, and number of classes as necessary.
- Add a text box to the map document and describe what affected your choice of classification method.
- Change elements such as the titles, colors, and legend to make a visually pleasing map.
- Save the results as **Exercise 2-2.mxd** in the \GIST2\MyExercises folder.

WHAT TO TURN IN

If you are working in a classroom setting with an instructor, you may be required to submit the maps you created in tutorial 2-2.

Turn in a printed map or screen capture of the following:

The three maps you created in tutorial 2-2

Exercise 2-2.mxd

Tutorial 2-2 review

Although you should strive for realism in your maps, there's no practical way to display all the quantities a field might contain. They must be simplified, or classified, to show a model of reality. There are many ways to classify data into groups, and you saw that not all will represent the data correctly. Some methods put all the features into one class, leaving a lot of blank space on the map. Others spread the data so evenly that no real pattern emerged.

2-1
2-2
2-3
2-4

The number of classes can also make a difference. Too many might confuse the viewer, and too few might make the map too simple to derive any meaningful answers. You also saw how to set these manually to conform to a more easily understood convention of showing years by decade, even though the automated classifications produced a nice result.

Choosing a classification all started with examining the distribution of the data. This gave you a chart of the values and pointed the way to a classification method that was more likely to produce the desired result.

STUDY QUESTIONS

1. What is the difference between an equal-interval and a quantile classification?
2. What does the frequency distribution diagram show?
3. Describe examples of outliers. How would you deal with them?

Other real-world examples

A quantile classification might be used to display a dataset as percentages of the whole. Making four classes with the same number of features in each class would result in breaks of 25 percent.

An equal-interval classification might be used to show individual age, grouping the values by 10 to represent 10 years.

A natural breaks classification could be used to classify a dataset of many lower-value houses, few medium-value houses, and many high-value houses. Because the dataset has two natural groupings, the classes should be set to highlight these.

Tutorial 2-3

Creating a map series

A single map showing population breakdown was sufficient for the previous exercise, but sometimes you need to look at data from multiple attribute fields. An effective way to do this is with a map series. Each map displays different attributes in a similar way, allowing the viewer to do analysis across the maps.

Learning objectives

- *Create multiple map displays*
- *Look for patterns*
- *Work with multiple attributes*

Preparation

- *Read pages 56–60 in* The ESRI Guide to GIS Analysis, Volume 1.

Introduction

In the Census 2000 data, there is a series of fields representing ethnic breakdown. Using the graduated colors, you can display only one of these features at a time. So it's easy to see concentrations of a particular ethnic group but difficult to see how that concentration compares to other ethnic groups.

One way to solve this problem is by creating a map series. You would make a new map for each ethnic group, symbolizing the quantities in a like manner. The viewer then examines all the maps side by side and makes comparisons of the data. One important thing to remember when doing a map series is that the data must be classified the same on all the maps in order for the viewer to easily make comparisons. If you showed Hispanics on one map with three classes, and Pacific Islanders on a second with five classes, the viewer could not compare the highest values of the two maps. They might be the same shade of a particular color, but they would represent different ranges of concentrations.

Scenario You're still interested in the investment opportunity presented by the grocery store chain, but you want to see if the idea would transfer to other ethnic groups. You'll create a map series showing different ethnic groups across the county and compare them. You are looking for another ethnic group that might lend itself to the development of a new chain of specialty grocery stores.

Data You can continue to use the Census 2000 block-group-level data. In the attribute table, there is a series of fields representing different groupings of counts that you can map.

Create the first map in the series

1 In ArcMap, open Tutorial 2-3.mxd. Once again you get the familiar map of Tarrant County containing the unsymbolized census data.

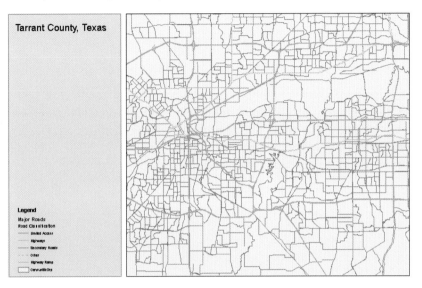

2-1
2-2
2-3
2-4

The goal is to create a series of maps displaying the various ethnic groups. When you do this, you want to make the maps have the same "look and feel" so that the viewer can easily compare the data across the series. You also want to make sure that the maps cover the same area at the same scale, and most importantly have the same ranges on the legend. On one ethnic group, you may get a maximum count of 2,000, while on another you may get a maximum count of 20,000. If both of these values are displayed with the same intensity of color for the top group, the viewer might make the false assumption that the total numbers of each group are similar.

For this series, you want to create maps showing the total population, the Hispanic population, and the black population. These three maps can be viewed adjacent to each other to give a good idea of how these groups are dispersed across the county.

Make the map of the total population first, then use the classification ranges for that dataset to display the other datasets. This will give you the best cross-map comparison.

2 Open the properties for CensusBlkGrp and go to the Symbology tab.

3 Select Quantities > Graduated colors and set the Value field to POP2000.

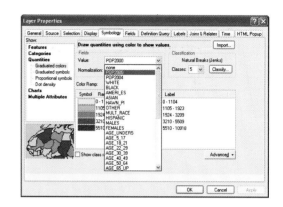

4 Accept the default classification, change the color ramp if you wish, and click OK.

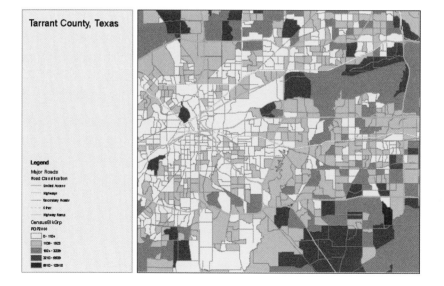

Next, you will change the title of the map to reflect the data shown.

5 Right-click the map title and select Properties. Change the title to **2000 Total Population** and click OK.

6 On the main menu, select Insert > Title. ArcMap will insert a title text block at the top of your layout.

7 In the Properties dialog box, change the title of the new text box to **Tarrant County, Texas.** Drag the text over to the left and place it under the map title. Don't worry if the text looks too big or if it isn't exactly centered under the other text. There are some tools you can use to fix that.

8 On the Draw menu at the bottom of the map document, click the Font Size drop-down box and select 20. If you aren't sure which is the Font Size box, point to the different icons on the menu with your mouse and read the ToolTip.

Next, you will fix the alignment.

9 Hold down the Shift key and select the map title. You should now have both text boxes selected.

10 Right-click the text and select Align > Align Center. Then move the text back to the top of the gray box. Another way to accomplish the alignment would be to use text formatting in the text-editing box. This would also allow you to do a multiple-row title with different fonts or text sizes. Read more about this by clicking the About Formatting Text button in the text properties box.

11 Print the map or export it as an image or PDF file.

Now you will duplicate the process to make the other two maps in the map series. First, you will copy the census data file so that you can symbolize it for the next field.

Create the second map in the series

1 Right-click the layer CensusBlkGrp and select Copy.

2 Right-click the data frame Tarrant County and select Paste Layer(s). The layer is pasted into the table of contents, but it looks just like the other layer and has the same name. Change the name of the layer to avoid confusion and the symbolization to reflect a different set of data. Remember that it is

important to use the same classification scale as the total population, so you will use the Import button in the symbology editor to match the other layer.

3 Open the layer properties of the copied dataset and go to the General tab. Type **Hispanic Population** as the name of the new layer.

2-1
2-2
2-3
2-4

4 Go to the Symbology tab and click the Import button.

5 There's only one choice of layers to import from, so click OK.

6 Change the Value Field to HISPANIC to use this other field for your symbolization, and click OK. Notice that the value field is now HISPANIC, and the ranges for the classification are the same as those for the total population. It would be fine to let ArcMap automatically set these ranges if you were just looking at one ethnic group, but in order to compare them they need to reference the same ranges.

7 When you are finished reviewing the settings, click OK.

8 Turn off the CensusBlkGrp layer.

9 Lastly, you need to change the map title. Double-click the map title to open the Properties dialog box and type **2000 Hispanic Population**. Click OK to accept the changes.

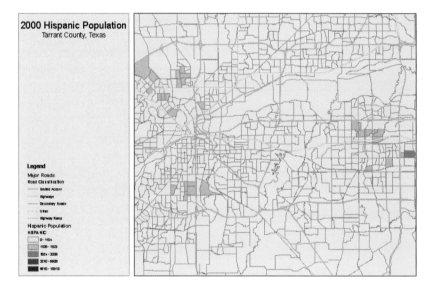

10 Print the map or export it to an image or PDF file. Look at both of these maps side by side. On the Hispanic population map, you don't see the darkest color showing a large population, but you still see some groupings. And you are using the same reference to compare the values.

11 Save your map document as **Tutorial 2-3.mxd** in the \GIST2\MyExercises folder. If you are not continuing on to the exercise, exit ArcMap.

YOUR TURN

Repeat the process and create a map using the field BLACK to show the population of blacks. Make sure to import the symbology and retain the same classification ranges. Change the title to read 2004 Black Population. When you are finished, print or export the map to an image or PDF file.

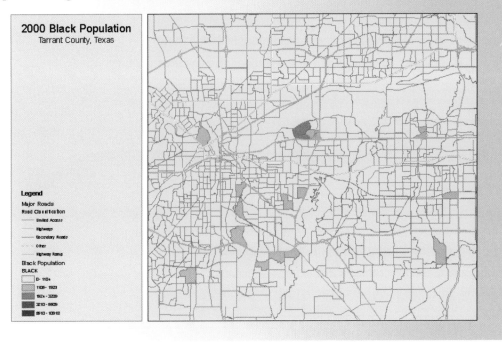

Exercise 2-3

The tutorial showed the process of making a map series showing ethnic breakdown in Tarrant County, Texas. The key was using the same classification values on all the maps so that comparisons can be made between them. The maps showed the count for various ethnic groups.

In this exercise, you will repeat the process showing concentration rather than the count.

- Continue using the map document you created in this tutorial or open Tutorial 2-3.mxd from the \GIST2\Maps folder.
- Symbolize CensusBlkGrp with graduated colors based on the HISPANIC value field, making sure to use the same ranges as before.
- Normalize the values by the total population using the POP2000 field.
- Insert a text box and describe the difference between using a count and using a concentration.
- Change elements such as the titles, colors, and legend to make a visually pleasing map.
- Save the results as **Exercise 2-3.mxd** in the \GIST2\MyExercises folder.

WHAT TO TURN IN

If you are working in a classroom setting with an instructor, you may be required to submit the maps you created in tutorial 2-3.

Turn in a printed map or screen capture image of the following:

The three maps you created in tutorial 2-3

Exercise 2-3.mxd

Tutorial 2-3 review

Since you wanted to compare the values across the different maps, it was important to be careful to use the same classification ranges. If you were only interested in each map individually, you could reset the classification ranges for each map. But in order to do cross-map comparisons, you had to use the same base classification ranges in each map.

2-1
2-2
2-3
2-4

This map series showed the differences between various data values captured at the same point in time. Another type of map series shows how a single data value changes over time. This topic is covered later in this book.

STUDY QUESTIONS

1. Why is it important to keep the classification ranges the same in a map series?
2. Would the same rules apply for classifying points, lines, or polygons?
3. Could you show several of the field values together on the same map?

Other real-world examples

The Texas Department of Parks and Wildlife may want to map bird counts done at each state park in a given month. In order to compare which parks have the highest concentrations of birds, it would make a map series showing each bird type normalized by the total bird count.

The Texas Department of Transportation may have data showing both auto and tractor traffic counts. Each may be used to create a concentration map. The two maps produced would have the same classification range values so that the numbers could be compared on the same scale.

Tutorial 2-4

Working with charts

You saw that a map series is helpful in comparing similar values across several maps. But the solid color fill prevented you from showing more than one field value on a map. Charts, however, will let you show several fields at the same time.

Learning objectives

- *Create charts*
- *Select fields for charts*
- *Compare field values*

Preparation

- *Read pages 61–62 in* The ESRI Guide to GIS Analysis, Volume 1.

Introduction

A chart can display several attributes at the same time and allow the user to quickly compare values against the same scale. The charts can easily be constructed from the attribute tables of your data and added to your map layouts. As you'll see in this tutorial, however, charts have visual limitations.

Creating charts on a large area map with a lot of small polygon areas may make the charts difficult to see, and the software will automatically remove some of the overlapping charts for clarity. Charts are best presented on a small area map where more detail can be shown. There are three types of charts available in ArcMap: a pie chart, a bar or column chart, and a stacked chart. With all, you will set the symbology method to Chart and select which fields are to be displayed in the chart. These fields will be used for visual comparison by the map reader, and they should have some basis for comparison.

With pie charts, the fields selected must represent the whole of what the values represent. For instance, if you are representing marital status, you can't just show married and divorced. There are more categories, and you would not be representing all the people surveyed. You would need to include categories such as single and widowed to represent all people. By looking at the pie wedges, the viewer can get a feel for the portion of the total each value represents. These do not lend themselves to showing values very effectively. By looking at two pie charts, you may see that one category has a larger pie wedge, but you cannot tell exact values.

Bar charts, on the other hand, do show amounts rather than the percentage of the total. Each value is shown as a bar in the chart and a comparison can be made between the heights of the bars. Stacked charts also show values, but display the columns on top of each other rather than side by side. The column heights are additive, with the total column height representing the total of all the field values.

Which chart style you use depends on if you want to show percentages (pie charts) or compare actual values (bar or stacked charts).

Scenario Continuing with the investment opportunity in ethnic-themed grocery stores, you found an area that seems ripe for expansion. The bankers are considering your loan, but they want to make sure that you'll have a chance at long-term success. It would be better if the houses in the area were primarily owner occupied so that each store can build a relationship with the customers and keep them coming back for a long time. If the houses in the area are mostly rental units, the customers may be more transient and not support the community efforts you are making. They may also not have as strong a customer loyalty as permanent residents. You also want to note the number of vacant houses, which may represent future customers who will take pride in their neighborhood.

Data Continue using Census 2000 block-group-level data. In the attribute table, there is a series of fields representing the occupancy status of the houses in the area that you'll be able to chart. Since the total of vacant, owner-occupied, and rental houses would represent the total number of houses, you can use pie charts to show the data.

Examine the attribute table for the census data

1 **In ArcMap, open Tutorial 2-4.mxd.** The map shows an area of great Hispanic concentration. Some ethnic-themed grocery stores are located near here, but the potential for expansion seems pretty high. Now, if you can just prove to the bank that the housing population here is stable, getting the loan will be no problem.

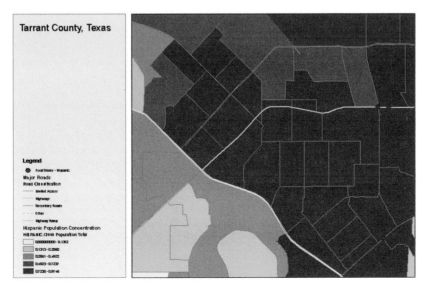

2 Right-click the layer Census2000 and open the attribute table. Notice the fields VACANT, OWNER_OCC, and RENTER_OCC. In order for these to be valid in a pie chart, the total of these values for each census block must represent the total number of housing units. To check that, choose one of the rows and add up the values for these three fields. They should equal the value of the field HSE_UNITS.

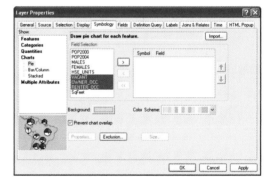

3 Close the attribute table and turn on the Census2000 layer.

Create a pie chart

1 Open the Properties dialog box for the Census2000 layer. On the General tab, change the layer name to **Housing Status.**

2 Go to the Symbology tab and select Charts > Pie in the Show window. Highlight the three fields for the chart—VACANT, OWNER_OCC, and RENTER_OCC—and click the ">" button. You can add these one at a time or hold down the Control button and select them all together.

Remember, the fields you select must represent the whole; and in this case they total the number of housing units. Once the fields are transferred over, you will set the colors for each pie segment. The Hispanic concentration layer you have already uses darker hued colors, so you should select brighter colors for contrast.

3 Right-click the symbol box for VACANT and select Big Sky Blue from the color palette.

4 Right-click the symbol box for OWNER_OCC and select Quetzel Green. Change the symbol color for RENTER_OCC to Aster Purple. Click Apply.

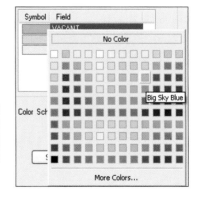

You will see that the charts have been drawn, but the layer is covering the Hispanic Population Concentration layer. To correct this, you will set the background of the Census2000 layer to clear.

5 **Click the color box next to Background and select Hollow. Click OK and OK again.** The map now displays the pie chart of housing status on top of the Hispanic concentration layer. By looking at the pie charts, you can see that the vacancy rate is low and that most of the census blocks have a high percentage of owner-occupied housing. Only a few areas have a larger percentage of renter-occupied housing than owner-occupied housing, so you will be able to show the bankers that the neighborhood has a stable customer base.

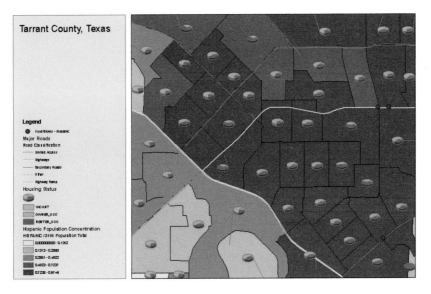

YOUR TURN

Experiment with other types of charts and map scales.

1 Change the chart type to Bar Chart and observe the differences. You may also want to try a stacked chart to see what effect it has on the map display. You may need to change the size to get the charts to be readable. When you are done, change the chart type back to Pie Charts.

2 Select Bookmarks > Zoom 2. What happens to the charts at this scale?

---

6 Save your map document as **Tutorial 2-4.mxd** in the \GIST2\MyExercises folder. If you are not continuing on to the exercise, exit ArcMap.

Exercise 2-4

So far we've shown an area of great interest for grocery store development by finding an area with a high Hispanic population. Then you demonstrated to the bankers that the housing status of the area is favorable—one with a high percentage of owner-occupied houses.

In this exercise, you will repeat the process of creating a chart to show the male versus female population for each census tract.

- Continue using the map document you created in this tutorial or open Tutorial 2-4.mxd from the \GIST2\Maps folder.
- Display chart symbols for male and female population; choose which style you feel would best display these values.
- Set appropriate colors for the chart.
- Change elements such as the titles, colors, and legend to make a visually pleasing map.
- Save the results as **Exercise 2-4.mxd** in the \GIST2\MyExercises folder.

WHAT TO TURN IN

If you are working in a classroom setting with an instructor, you may be required to submit the maps you created in tutorial 2-4.

Turn in a printed map or screen capture of the following:

Tutorial 2-4.mxd

Exercise 2-4.mxd

Tutorial 2-4 review

You looked at several types of charts that can be created with ArcMap. The pie chart shows each value as a percentage of the whole, provided that all the component fields equal the whole. It does not, however, give you a good feel for amounts.

Bar charts and stacked charts give a better display of actual values from the fields. Bar charts let you compare field values to each other, and stacked charts let you also compare the totals for an area with the totals for other areas by looking at the height of the stack. They present other challenges in displaying them to read clearly.

Charts in general are an excellent way to show attribute data overlaid on other data. By setting the background of the chart layer to hollow, you could see underlying analysis display as well as the information from the charts. You do need to be aware of scale as charts tend to clutter and overlap on small area maps. Another caveat with charts is not to include too many fields. The tutorial's pie chart with 3 fields was easy to read, but imagine how difficult it might be to read a chart with 14 fields. A bar chart with 14 fields would be very wide and may not fit inside the polygon it is representing.

2-1
2-2
2-3
2-4

STUDY QUESTIONS

What are the advantages and disadvantages of each chart style in regard to the following issues?

- Scale
- Values versus percentages
- Number of fields
- Other data layers

Other real-world examples

The police department has crime statistics for each police district in the city and wants to show a comparison of last year's crime rate and this year's crime rate. A map with the police districts color shaded in the background might have a bar chart shown in each district with both years' crime rates. A comparison of the values could be made by looking at the chart, and the activity in one district could be compared to another district by comparing the sizes of the bars.

The city secretary has compiled the election results and wants to display the vote totals for each precinct. The totals could be displayed with a map showing the voting precincts in the background and a pie chart with each candidate's total represented by a wedge in the pie.

The National Energy Commission wants to display data showing how much of each state's energy comes from oil, natural gas, or nuclear power. It might create a map of the United States with a bar chart over each state displaying the totals. The sizes of the bars could also be compared between states to determine which state uses the most energy.

3

Mapping density

Values associated with an area can be shown simply as their value, but they may have a greater impact if shown in comparison to the area of a feature. The relationship between polygon feature values and the area of the polygon is called density. Using density, factoring in the impact of area, makes comparisons across features more precise. No special symbology is needed to show density in most cases. A simple color ramp from light to dark, with dark showing the highest density, is often sufficient. Other instances may use a special symbology called a dot density, or a method of showing density with raster datasets called a density surface.

Tutorial 3-1

Displaying density for analysis

Data summarized by area, such as census data, is often displayed as a straight value or as a percentage of the total, as was demonstrated in the previous tutorial. These values have no relationship to the size of the area they represent. Comparison across polygons of different sizes is difficult until you factor in area, creating a density value.

Learning objectives

- *Create density values*
- *Compare data across polygons of different sizes*

Preparation

- *Read pages 60–75 in* The ESRI Guide to GIS Analysis, Volume 1.

Introduction

In the previous tutorial, you looked at data shown by the amount contained in an area. It was easy to compare values across polygons as percentages or totals, but the data may be giving a false picture of the values.

An amount contained in a large area does not have the same significance of an equal amount contained in a small area. Imagine if you compared 50 people in 100 acres of land to 50 people in 1 acre of land. The people per acre (amount divided by area) tells a different story than just the count, especially if the polygons vary greatly in size.

Any value divided by the measured area it represents is called a density. It is common to hear people refer to people per square mile, value per square foot, or crop yield per acre. Each of these is a value divided by an area measurement. In fact, you can substitute the words "divided by" when you hear the word "per." Note that the area units can be different for each density calculation, so it is important to display the area units on your map.

When density is shown on a map, area is removed as a factor in comparing values. A large residential subdivision with three housing units per acre is no denser than a small subdivision with three housing units per acre.

Scenario The city planner wants to see a map of population totals and population density for the year 2000 in people per square mile. The first map is simply the population value field symbolized with graduated colors. The second map will divide the value by the area to display density, or people per area value. "Per" can mean "divided by" for your purposes, and if you recall, the word "normalize" also means "divided by." You should be able to use the normalization field in ArcMap to easily create a density map.

Data　The data is the Census 2000 block-group-level data. This particular set comes from the ESRI Data & Maps media kit that comes with ArcGIS software. It contains fields for the total population in 2000 and 2004. The area needed is cut out for you to use, but the analysis could be repeated for any area with the data from the media kit.

The other set of data is the street network data to give the map some context. It also comes from the ESRI Data & Maps media.

3-1
3-2
3-3

Map population density by census block

1 **In ArcMap, open Tutorial 3-1.mxd.** The map shows the Dallas/Fort Worth metropolitan service area, a four-county region with a population of about 5 million people. You're interested in seeing the total 2000 population by census block group, and a population density of the same data. First, you need to make a copy of the CensusBlkGrp layer, then you'll symbolize the two layers to display the desired data.

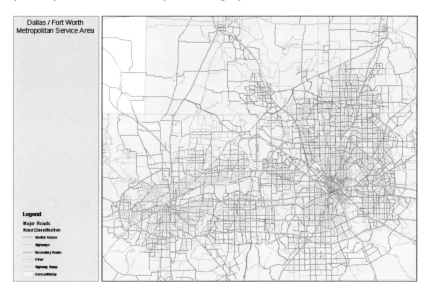

2 **Right-click the CensusBlkGrp layer and select Copy.**

3 Right-click the Tarrant County data frame and select Paste Layer(s). Turn the copied layer off for now.

Now change the layer name and symbology properties for the first layer. This is what will display in the legend, so it should be very descriptive of the data it will represent.

4 Right-click the visible CensusBlkGrp layer and select Properties. Go to the General tab and type **2000 Population** for the layer name. Click Apply.

5 Go to the Symbology tab and select Quantities > Graduated colors in the Show window.

6 Set the Value field to POP2000. Accept the default classification and colors. Click Apply.

Examine the change in the map display. This is a very quick and simple display of population. The contrast looks good, but seeing the population count numbers in the legend may not mean a lot when viewed at this scale. You really only want to know relative population, so you'll change the labels to read Low, Medium, and High.

7 Click in the Label column for the lowest value and replace it with **Low.** Click the next label and press the delete key, removing the label altogether. Make the middle label **Medium,** leave the next label blank, and name the highest value label **High.** Click OK. The map now displays the total population for each census block group, and your simplified legend makes the map a little easier to read. Note where the darkest areas are located, and you'll see later if these are in fact the areas of the highest population density.

8 Print this map or create an image file of it.

Next you want to use the copy of the CensusBlkgGrp layer to show population density, or people per area. You'll use POP2000 as the value field again, and you can use the normalization function to automatically divide by area. The city planner wanted to see the results in people per square mile, so you'll need a field that contains the square mileage for each census block group. First add a new field to the table, then you'll calculate its values in square miles.

Map population density per square mile

1 Right-click the CensusBlkGrp layer and open the attribute table. Click the Table Options drop-down menu and select Add Field.

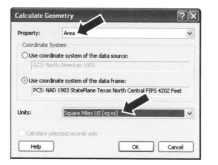

2 Enter **SqMiles** as the new field name, and then set its type to Float in the Type drop-down list. Then click OK.

3 Right-click the new SqMiles field and select Calculate Geometry.

4 Click OK on the warning message. In the Calculate Geometry dialog box, make sure the Property is set to Area, and then use the drop-down menu to set the Units field to Square Miles US. Then click OK.

5 Close the attribute table.

6 Turn on the CensusBlkGrp layer, open its properties, and go to the General tab. Change the name to **2000 Population Density** and click Apply.

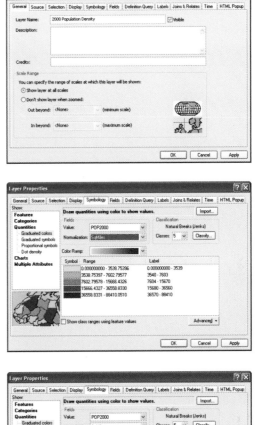

7 Go to the Symbology tab and select Quantities > Graduated colors in the Show window. Set the Value field to POP2000 and the Normalization field to SqMiles.

Once again you should simplify the legend so that the viewer is getting a general feel for the data, and not necessarily concentrating on individual values.

8 Set the labels to match the 2000 Population layer, changing them to **Low, Medium,** and **High.** When everything matches the image, click OK. This process added a field, calculated an area, and used that area to set up a display of density. Changing the labels in the legend made the map easier to read and suitable for a general audience.

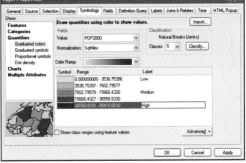

3-1

3-2

3-3

9   Turn off the 2000 Population layer. The resulting map is very different from the previous map. What you saw as large, dark polygons representing lots of people turned out to be medium-to-low density. Now it becomes easy to see the concentrations of people.

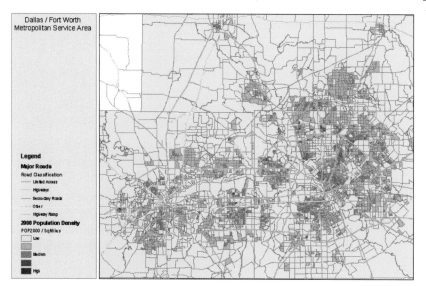

10   Print this map or create an image file of it.

YOUR TURN

Create two more maps: one showing the 2004 total population and one showing the 2004 population per square mile using the POP2004 field in the CensusBlkGrp layer. This will give the city planner two sets of data to compare and will show how the population and the population density in the area have changed over time.

These printed maps nicely display how things have changed from 2000 to 2004, but it may be hard to compare two specific areas. Flipping one sheet over the other is not very efficient. Now you will look at an ArcMap tool from the Effects toolbar that will let you do this with great ease.

Compare maps using the Swipe tool

1 Turn on the layers 2000 Population and 2004 Population. Turn off the two population density layers. On the main menu, select Customize > Toolbars > Effects.

2 Since Effects tools work only in data view, on the main menu select View > Data View.

3-1

3-2

3-3

3 Make sure the 2000 Population layer is above the 2004 Population layer in the table of contents by dragging it up higher in the list.

4 Set the Layer in the Effects toolbar to 2000 Population.

5 Select the Swipe Layer tool on the Effects toolbar.

6 Move into the right center of the map area, then click and hold the left mouse button. It may take 10 or 20 seconds for the map to be loaded into the Swipe Layer tool memory. When the hourglass disappears, move the cursor to the left and notice that the 2000 Population layer is being replaced as you move the tool across the map.

The Swipe Layer tool is very effective for seeing how features have changed between two layers. Grouped layers can also be set as a swipe layer, allowing you to swipe sets of data. Once you have finished examining this data, set up the Effects tool to look at the change between the 2000 Population Density and the 2004 Population Density.

7 Save your map document as **Tutorial 3-1.mxd** in the \GIST2\MyExercises folder. If you are not continuing on to the exercise, exit ArcMap.

YOUR TURN

Turn off the two population layers and turn on the two population density layers. Set up the Swipe Layer tool for these layers and examine the changes in the four-year span.

Exercise 3-1

The tutorial showed how to map census values as both totals and densities. The values used were population counts, but other data is also useful when mapped as a density.

The city planner also wants to show the density of housing units, expressed in households per acre. The field HOUSEHOLDS in the CensusBlkGrp layer has the number of households in each census block group.

3-1
3-2
3-3

- Continue using the map document you created in this tutorial or open Tutorial 3-1.mxd from the \GIST2\Maps folder.
- Add a field to the CensusBlkGrp attribute table and calculate the acreage for each polygon.
- Change the name of CensusBlkGrp layer and symbolize the households.
- Normalize by acreage.
- Change elements such as the titles, colors, and legend to make a visually pleasing map.
- Save the results as **Exercise 3-1.mxd** in the \GIST2\MyExercises folder.

WHAT TO TURN IN

If you are working in a classroom setting with an instructor, you may be required to submit the maps you created in tutorial 3-1.

Turn in a printed map or screen capture of the following:

The four maps in Tutorial 3-1.mxd

Exercise 3-1.mxd

Tutorial 3-1 review

In this exercise you saw how density displays data differently from straight counts or totals. By dividing the total value by the area it represents, densities let you do comparisons between dissimilar areas by showing the data as a concentration. Since the normalization feature lets you specify a field to divide into the value field, you can easily set up density displays using any field.

While fields that show counts or totals summarized for an area benefit from being displayed as densities, there are others that are not as suitable. Fields that display calculated values, such as averages or percentages, are not suitable for density maps. These fields have already been divided by another value, so dividing by area doesn't make sense. A field representing a percent would be the value divided by the total. Showing this as a density would be the value divided by the total, divided by the area. A map of "Percent Renter Occupied per Square Mile" has no real meaning.

STUDY QUESTIONS

1. Can any field be used for density mapping?
2. Why does using densities or concentrations allow a better comparison between values that have been summarized by area?

Other real-world examples

A police department may summarize data by police beat and display the density of a particular crime. This would help it determine how many officers to assign to each beat.

A political campaign might take voter registration rolls and summarize them by county. Then it could display the areas of concentration of registered voters to target its efforts.

Census data is one of the most common things to display as a density. Any of the data fields representing totals or counts could be displayed as a density, such as people per square mile or concentrations of high school graduates.

Tutorial 3-2

Creating dot density maps

One drawback of density maps that use shaded polygons or a density surface is that only one value can be mapped at a time. The dot density map allows you to display the density using a pattern that can be overlaid with other data.

Learning objectives

- *Create a dot density map*
- *Overlay density data*
- *Analyze patterns*

Preparation

- *Read pages 76–77 in* The ESRI Guide to GIS Analysis, Volume 1.

Introduction

In the previous tutorial, you displayed data divided by the area it represented. This type of density mapping works very well in most situations, but causes problems if you want to display multiple values together. The solution is to use a dot density map.

When creating a dot density map, you select a value field, a dot size, and a number of units that the dot will represent. ArcMap reads each value and calculates how many dots to display in the polygon area. For example, if the field value is 1,200 and the dot value is 20, then ArcMap will randomly display 60 dots in the polygon.

It is important to note that the dots are placed randomly. This works well in small areas, but in larger areas the dots may be randomly grouped in a region of the polygon and suggest that the dots represent some sort of data grouping.

Another important factor in making a dot density map is the dot size. This is set in conjunction with the dot value to determine how big the dots should be. Too big and the dots will overlap and obscure the map; too small and they won't be noticeable against the background. You may need to experiment with the dot size and dot value to get the best possible display for the scale of map you are producing.

Scenario The parks director wants to build a dog park in Oleander and needs help finding the right place. He feels that the park would be used mostly by apartment dwellers who do not have a large yard in which the dogs can play. So you are going to look for an area with a large population density as well as a concentration of rental units.

Data The data is the Census 2000 block-group-level data. This particular set comes from the ESRI Data & Maps media kit that comes with ArcGIS. It contains the field POP2000, which represents the population count in 2000, and RENTER_OCC, which represents the number of occupied rental units. The area needed is cut out for you to use, but the analysis could be repeated for any area with the data from the media kit.

The other set of data is the street network data to give the map some reference. It also comes from the ESRI Data & Maps media.

Map the density of rental units

1 In ArcMap, open Tutorial 3-2.mxd. The map shows the City of Oleander with the 2000 Population Density already color-shaded. You need to add the concentration of rental units in such a way as to not obscure the population data. To do this, you'll set a dot density classification and show the patterns on top of the population layer.

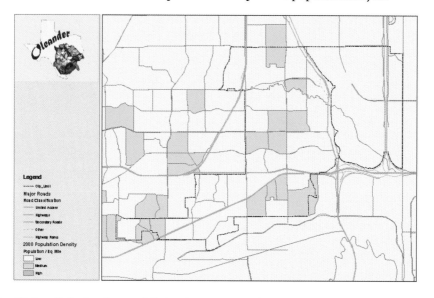

2 Right-click the CensusBlkGrp layer, open the properties, and go to the General tab. Type in **Concentration of Rental Units** for the title and check the Visible box to make the layer visible. Click Apply.

3 Go to the Symbology tab and set the Show window to Quantities > Dot density. Notice that the dialog box will let you add multiple fields to the Symbol list for dot density. Just like making charts, the list should contain values that can be compared on the same scale. You can use as many fields as you like, but too many will make the map difficult to interpret. Having more

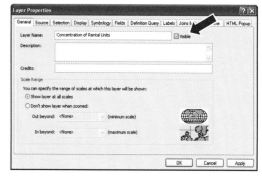

than one symbol value also adds a degree of difficulty to setting the dot size and scale. There will be more dots per polygon, and they may begin to interfere with each other.

4 Select the RENTER_OCC field and click the ">" button to add it to the Symbol list. Click Apply and note the change in the map. Try different colors, dot sizes, and dot values to get the map to look the way you feel is the best. Click Apply to check the results of your settings. When you are satisfied, click OK.

3-1

3-2

3-3

The resulting map shows the concentration of rental units as dot patterns, with the underlying population density still visible. Click the Refresh View button at the bottom of your layout a few times and notice what happens to the dot distribution—it changes each time. A quick visual analysis shows a place in the southwestern part of town that would be perfect. It's got a high population density and a concentration of rental units, and is located on a major thoroughfare. This will be enough to investigate availability of land and perhaps do a survey of what amenities might be desirable.

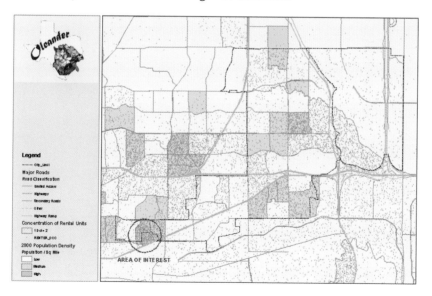

There are two options for the dot density classification that can be used to add more control to the location of the dots. These are accessed by clicking the Properties button on the Symbology tab. The first option (Dots) allows you to lock the dots in place and keep them from moving around when you refresh the map. While their placement is still random, it will let you keep a consistent look of the map for things such as a map series.

The second option (Masking) can cause the dots to be displayed with respect to another polygon feature class. The first choice will cause the dots not to be displayed within the noted polygons. This may be useful if there are large lakes or ponds in the area. You wouldn't want to display a density of people over a lake. The second choice will display the polygons only in the referenced area. Perhaps you want to show data from the Census layer, but only within your city limits. Note that the restricting layer does not have to be the one for which you set the dot density classification.

5 Save your map document as **Tutorial 3-2.mxd** in the \GIST2\MyExercises folder. If you are not continuing on to the exercise, exit ArcMap.

Exercise 3-2

The tutorial showed how to set up a dot density and display the density data over another color-shaded dataset. For this exercise, you will create another dot density map showing two fields as values.

The parks director wants to see the male-to-female comparison in the area of interest. You'll try to show that data as a dot density map with the MALE and FEMALE fields from the CensusBlkGrp layer.

Continue using the map document you created in this tutorial or open Tutorial 3-2.mxd from the \GIST2\Maps folder.

3-1
3-2
3-3

- Change the name of CensusBlkGrp and make it visible.
- Zoom to the bookmark Area of Interest.
- Set the symbology to Dot Density and use both the MALE and FEMALE fields.
- Change the dot color, dot size, and dot value to make the data read well.
- Change elements such as the titles, colors, and legend to make a visually pleasing map.
- Save the results as **Exercise 3-2.mxd** in the \GIST2\MyExercises folder.

WHAT TO TURN IN

If you are working in a classroom setting with an instructor, you may be required to submit the maps you created in tutorial 3-2.

Turn in a printed map or screen capture of the following:

Tutorial 3-2.mxd

Exercise 3-2.mxd

Tutorial 3-2 review

In this exercise you saw how density can be displayed as dots. You were able to control the size and number of dots displayed. ArcMap, however, determines the actual random placement of the dots. Because of this, you must only use the patterns the dots create as a general view of density and not read them as grouping or clustering. Every time you redraw the map, the random location of the dots changes, potentially giving a different reading of the map. These can be locked down, but remember that they do not represent groupings, only a random scattering of dots.

Dot density maps can be created with several value fields, but you saw how this can make the map hard to read and interpret. Care must be taken in setting the dot value, size, and color.

One advantage of using dot density is that you were able to display a set of data on top of another symbolized layer. In this case, the dots were displayed over total population density so that the viewer could look at both the dot patterns and the color shading for visual analysis.

STUDY QUESTIONS

1. When would it be more practical to use dot density maps than charts?
2. Is dot density a true density?
3. A dot density map of the United States is created using each state as one polygon and the total population as the value field. What is the problem with this scenario? What would the map look like? Could the patterns lead to false conclusions?

Other real-world examples

A police department may want to show concentration of crime without revealing actual victim locations. It might summarize crime data by police beat, then create a dot density map of the totals to show general concentrations. The random assignment of dots would not represent real crime locations, only symbolize that crime had occurred.

A political campaign might want to display the number of registered voters as a dot density on top of the total population of voting age per precinct. Areas with a high eligible population but low registration might be a good target area for recruitment.

Census data lends itself so well to density mapping, and dot density allows the overlaying of several types of data on a single map. With the total population as the backdrop, interesting maps could be created showing housing.

Tutorial 3-3

Creating a density surface

A density map was easy to create using the data summarized by area. The polygons made it easy to create solid shade or dots to represent density. If the available data is point data, however, it is not as easy to show density. A density surface will take the points and create a continuous set of values to display density in a raster format.

Learning objectives

- *Create density surfaces*
- *Work with raster data*

Preparation

- *Read pages 78–85 in* The ESRI Guide to GIS Analysis, Volume 1.

Introduction

The previous tutorial worked with data that was already summarized by area. It was fairly easy to color-shade the polygons to represent density, or to make a dot density map. With data that is not summarized by area, you must create a density surface in order to see the data projected as a density. A density surface is a raster image created from points, making a set of discrete features into a continuous phenomenon dataset.

There are several parameters to set when calculating a density surface. The first is cell size, which equates to pixel size in the output data. This can determine how fine or coarse the patterns will appear. A small cell size will create a finer-resolution raster image but will take longer to process. Conversely, a large cell size will create a coarser image but process quickly. ArcMap will suggest values for you based on the area you are working with.

The second parameter that you will deal with is the search radius. When ArcMap creates the density surface, it will split your map extent into pixels of the size you indicate. Then it will count all the features that fall within the search radius of each feature, divide that total by the area of that feature's search area, and create a pixel with that value. Then it moves on to the next pixel, working its way through the entire dataset. A large search radius will take longer to process and will give the surface a more generalized look. Small variations in the data may be missed because too many features are falling within each search radius. A small search radius will reflect more local variation, with fewer features having an effect on each pixel.

A density surface may reflect only the features and how they are grouped into the search area, or it may include a weight factor. This will give some features a more important role in the calculation of the density. For instance, a density surface made using only business locations might show one set of data grouping. But a density surface weighted by the number of employees will create groupings nearer to businesses with a lot of employees.

Scenario The parks director was impressed with your analysis of suitable sites for the dog park and has come to you again with another request. Oleander is recognized as "Tree City USA" by the National Arbor Day Foundation, and the director wants you to do a tree inventory and tree-density map of all the city parks. An intern with a Global Positioning System (GPS) unit spent all last summer mapping trees. You now have the data collected and need to make a tree-density map. Since you are starting with point data, you'll need to make a density surface to demonstrate tree concentrations. You will also use the measured canopy size as a weight factor, making the larger trees have a stronger effect on the calculation.

Data The land-use data used to locate all the city parks is the city's cadastre file derived from recorded plats. Each parcel of land has a land-use code, as you may recall from exercise 1-1.

The tree data was collected by an intern from Texas A&M University's School of Forestry, who visited every park in the city and recorded the longitude and latitude of every tree. The intern also built a data collection menu to collect data such as tree type, diameter, various condition measurements, and estimated canopy width.

The other set of data is the street network data to give the map some context. It comes from the ESRI Data & Maps media.

Map tree density

1 In ArcMap, open Tutorial 3-3.mxd. The map is zoomed in to the central part of the city, where a tree survey has been completed. The areas outlined in green are city parks, and the brown dots are tree locations. You may think that this reads well as density, similar to a dot density map, but there are a few things you'll get from creating a density surface. First, you'll get an output file that represents area, not discrete points. Plus you'll be able to weight the impact of the trees on the density calculation based on their canopy size.

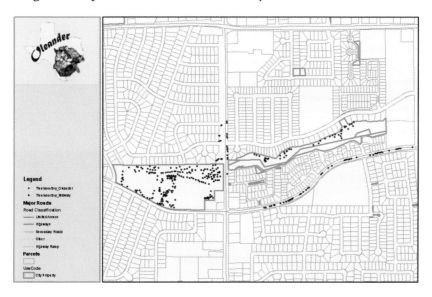

This tutorial uses the ArcGIS Spatial Analyst extension. You'll need to activate the Spatial Analyst extension in this ArcMap session. If you are unsure if you have this extension, select Customize > Extensions on the main menu and look for Spatial Analyst. Check it if it is not already checked. If it does not appear here, ask your system administrator for access.

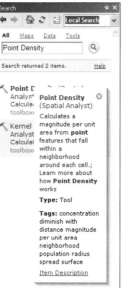

2 Click the Search tab on the right side of the map document window and type **Point Density** into the search window. Then click the Search button 🔍 to find the tool. Pausing over a tool name will show a description of the tool. Click the Point Density tool to run it.

3 In the Point Density dialog box, set Input point features to TreeInventory_Midway and the Output raster to Midway_Trees in the \MyAnswers\MyData.gdb geodatabase. **Note:** The raster output name is limited to 13 characters with no spaces.

The next values you will set are the search radius and the cell size. The Point Density tool has already analyzed your data and map extent and has suggested a suitable set of numbers. To get a finer result, you could decrease these values by about 20 percent; and to get a coarser result, increase them by the same factor.

4 Click the input box for Radius and type **53**. Replace the Output cell size with **6**. When your dialog box matches the graphic to the right, click OK.

5   Take a look at the tree locations overlaid on the new density surface to get an idea of how the two sets of data relate. When you are done, turn off the Tree Inventory layer to get the final map. The parks director will be able to show this map to the park board and help it plan the tree planting efforts for next year. You'll also be able to use this for area overlays in future GIS analysis such as showing the effect on the existing tree density of future playgrounds or athletic fields.

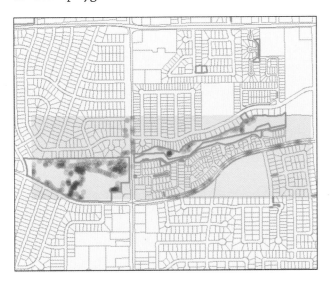

YOUR TURN

Move to the bookmark Oleander Park and repeat the density calculation for the Tree Inventory–Oleander layer. Eventually, the director will want this done for every city park.

6   Save your map document as **Tutorial 3-3.mxd** in the \GIST2\MyExercises folder. If you are not continuing on to the exercise, exit ArcMap.

Exercise 3-3

The tutorial showed how to map densities from point data. The points were used to create a density surface, and the result was a continuous phenomenon dataset.

Now the parks director would like you to do a more detailed study of tree density. He would like to see the density of elms and crape myrtles on two more maps.

- Continue using the map document you created in this tutorial or open Tutorial 3-3.mxd from the \GIST2\Maps folder.
- Open the properties of the Tree Inventory layer and set up a definition query to show only the elms.
- Create a point density raster of the resulting set.
- Change the definition query to crape myrtle and create a new point density raster.
- Make maps of each of these point density rasters.
- Change elements such as the titles, colors, and legends to make visually pleasing maps.
- Print or export images of the two maps.
- Save the results as **Exercise 3-3.mxd** in the \GIST2\MyExercises folder.

WHAT TO TURN IN

If you are working in a classroom setting with an instructor, you may be required to submit the maps you created in tutorial 3-3.

Turn in a printed map or screen capture of the following:

The two map layers you created in Tutorial 3-3.mxd

The two map layers you created in Exercise 3-3.mxd

Tutorial 3-3 review

The Point Density tool is used to make a continuous phenomenon dataset from discrete features. The surface can be used for further analysis with the data representing the values interpolated across a large area.

The tool creates a raster dataset with the specified cell size, then evaluates each cell for features that fall in and around it. Once each cell is given a value, a colored raster image displays densities or concentrations.

3-1
3-2
3-3

Care must be taken in setting the cell size and search area. Setting these values high will create a coarser map but process quickly. Setting them low will take more computer time but yield a finer result.

STUDY QUESTION

1. What are the three types of density mapping discussed in chapter 4 of *The ESRI Guide to GIS Analysis, Volume 1*, and when should each be used?

Other real-world examples

The local weather reporter gets rainfall and temperature readings based on discrete measuring stations. The reporter then takes the point data and creates a density surface showing a continuous phenomenon of temperature and rainfall.

A fire department may take point locations of calls for service and create a density surface showing the concentration of calls.

A wildlife researcher might take point data of animal observations and create a surface density to show animal density. This takes what are known to be limited sample data and interpolates it across a larger area.

geographic information systems geographic information systems geographic info
hic information systems geographic information systems geographic info
hic information systems geographic information systems geographic info
hic information systems geographic information systems geographic info
hic information sys geographic information systems geographic info
hic information sy formation systems geographic info
hic information sys ormation systems geographic info
geographic infor tion systems geographic info
graphic info
phic info
geographic info
on systems geographic info
information systems geographic info
hic information sys geographic information systems geographic info
hic information sys geographic information systems geographic info
hic information sys geographic information systems geographic info
hic information systems geographic information systems geographic info
hic information systems geographic information systems geographic info

4

Finding what's inside

GIS data deals with relationships among datasets, allowing for many analyses to be done using these relationships. One of the most basic is the concept of data overlays. Quite simply, does one set of data overlay another? This might be used simply as a selection process, with analysis done against the results. More complex processes might involve features that are only partially inside the overlay data. Analysis against the selection here cannot deal with the whole values for these features, but use proportions instead.

Tutorial 4-1

Overlaying datasets for analysis

Finding features inside a region is a powerful analysis tool. The region can be a graphic drawn by hand, or a feature that exists in another dataset. Once the features are selected, summations and comparisons can be done with the attribute information.

Learning objectives

- *Create a visual overlay*
- *Select features*
- *Obtain field summaries*

Preparation

- *Read pages 87–104 in* The ESRI Guide to GIS Analysis, Volume 1.

Introduction

A very powerful concept of spatial analysis is to decide if one set of features is inside another. For example, you might want to know if your house is in a floodplain, an earthquake zone, or in a good school district. To do this with GIS, a polygon region must be defined, then you can search for points, lines, or polygons that fall within those boundaries.

The boundaries used for an inside–outside analysis can be created as a graphic in ArcMap or may exist as a feature in another dataset. The graphics are drawn using the standard drawing tools in the layout view and can be used to do simple selections.

Using features from another layer for your selection is a little more complex but a lot more powerful. The desired boundary must be selected and a select-by-location query must be built. Conveniently, the query you build can be easily repeated for other regions. For example, to get a count of houses in a fire district, you would select one of the district boundaries from the Fire Districts feature class and use it in a select-by-location query against the residential parcel data. To repeat this for another district, just change the selected district and repeat the select-by-location query.

Once the features are selected, there are many things that you could do with their associated data. You could get a count of the features. The example above is a simple answer of how many houses fall within a fire district. You may also want to perform a summary operation against the data. This could be the total value of the houses or the average water usage.

Boundaries can also be created by buffering features, but since those deal with setting a distance from an existing feature, they will be covered in the "proximity analysis" tutorials.

Scenario   The city manager has outlined a large area of land on a map that represents the gateway to the city. The region has many shops, some vacant land, and an apartment complex. You need to select the property within that region and provide some summary statistics such as how many parcels and dwelling units are there, what the total area is, and how much of each of the land-use categories falls within the region.

Whether the city council vote is favorable, the city will use its parks department staff and plants from the city's greenhouse to implement a beautification project in the area. The city council meets tonight, and it has asked for a briefing on the project.

Data   The dataset you will be working with is the cadastre layer for Oleander. It represents platted and unplatted property and contains a field for land-use codes (UseCode). The field DU represents the number of dwelling units per parcel, and you'll use that for a housing count. You'll also use the Shape_Area field provided automatically by ArcMap for the area calculations.

4-1
4-2

Select features using a polygon

1 In ArcMap, open Tutorial 4-1.mxd. On the map is a region outlined in red with a gray crosshatch. This is the area the city's parks department is interested in beautifying.

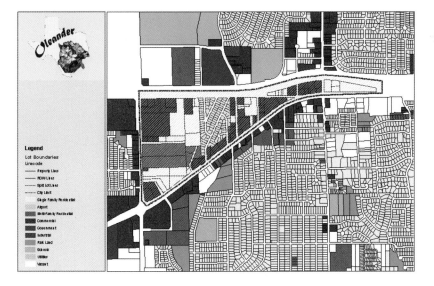

Do a preliminary investigation by looking at the map. Basic visual analysis will show that inside the region is some vacant land, some commercial land, some multifamily housing, and some single-family dwellings. So the city manager already has a general idea of land use inside the region and can give the city council a briefing while you take more time to complete the work. This is a simple but effective preliminary analysis.

To get the rest of the information, you'll need to select the parcels that fall within the region. First try doing a simple selection with a user-defined area.

2 Click the List By Selection tab at the top of the table of contents.

3 Click the icon to make Land Use the only selectable layer.

4 On the Tools toolbar, click the Select Features tool drop-down and choose Select by Polygon. The Select by Polygon tool will let you draw a polygon around the area you want to select freehand, and select whatever the selectable layers are. Always make sure to check what layers are selectable when you do this or you could have a mess afterward.

5 Using the selection tool, draw a polygon that follows the outline of the shaded region. Click to insert a vertex. Double-click the last point to finish. If you make a mistake, click the Undo button ↰ on the Standard toolbar and start over. The desired parcels are selected. You can now work with them to get the counts and summaries that you need. This method of selection worked pretty well, but there can be some problems with it.

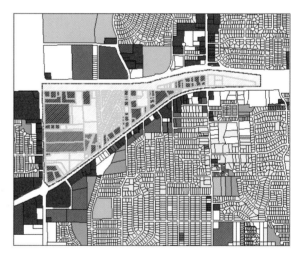

6 On the main menu, click Selection > Clear Selected Features. This can also be done by clicking the Clear Selected Features button ☒ on the Tools toolbar. Well, there's one of the problems. This selection method is not easy to repeat. Your task called for doing a summary, and now the selected set is cleared. Next you'll try another selection method. This method involves drawing a graphic of the area you want, then using it for the selection.

Select features using graphics

1 The Select By Graphics function only works in the Data View window. On the main menu, select View > Data View. Click the Select Elements tool ⬉ on the Tools toolbar.

2 In the map area, select the shaded graphic that was drawn by the city manager. It's hard to see on the map, but a rectangle that encompasses the shaded area is selected, as evidenced by a blue dotted line and selection points.

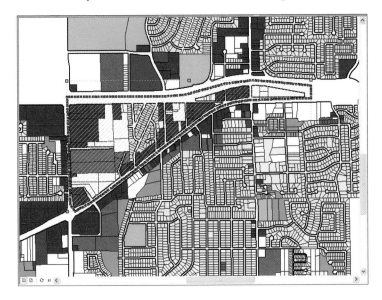

4-1
4-2

3 Click Selection > Select By Graphics from the main menu. All of the features inside the region of interest are again selected.

This graphic element surrounds the area of interest. So if the selected features are accidentally cleared again, they can be selected again very quickly using this graphic. If the graphic element had not already existed, you could have drawn one using the New Polygon tool on the Draw toolbar. These can also be symbolized in the same way that regular data features are symbolized.

Now you can proceed with the counts and summaries that are required for this analysis task. The first step is to get a count of how many parcels are involved in the project.

Obtain counts and summaries for features within an area

1 Go to the List By Selection tab at the top of the table of contents. The number appearing in parentheses next to the Land Use layer is how many features are currently selected. Looks like there are 202 parcels in the redevelopment region. That was easy.

Next you need to see how many dwelling units are in the region. The field DU has a value representing the number of dwelling units on each parcel. Vacant and commercial lots have a value of 0, lots containing single-family houses have a value of 1, and those with multifamily dwellings reflect how many are on each parcel.

2 Open the attribute table for the Land Use layer from the Selection tab by right-clicking the layer and selecting Open Table Showing Selected Features.

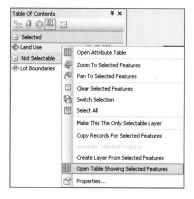

The table opens to display selected features. You can verify the number by checking the bottom of the table window. Yes, 202. Now get the total dwelling units.

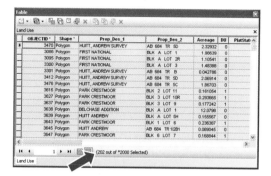

3 Right-click the field name DU and select Statistics. The resulting window displays the count, minimum, maximum, sum, mean, and standard deviation. Once again, you can verify the count at 202 and see that the total number of dwelling units is 633.

You also need to get a total area of all the parcels. This is done using the statistics feature again.

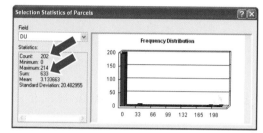

4 Close the Statistics window. Right-click the field Acreage and select Statistics. The resulting window shows all the stats of the field, including total acreage.

The total acreage is 152.1 acres. The city manager needed just one more piece of information: the breakdown of area for

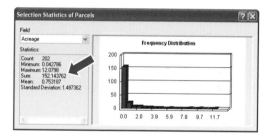

each land use. This analysis process so far has only produced the overall total. To get the breakdown, you'll use the Summarize command. This will take the selected features and a selected field, then report back each unique value of that field. In addition, you can add summary statistics to the process and get totals, averages, or standard deviations on the other fields in the table.

You can do this using the UseCode field. The Summarize command will list each unique value in the Land Use field and allow you to add a command to get the total acreage for each category.

5 Close the Statistics dialog box. Right-click the UseCode field and select Summarize.

6 In the Summarize dialog box, click the plus sign next to Acreage and check Sum. This will give the total acreage for each unique value of UseCode in the table. Check the box next to "Summarize on the selected records only," if necessary. Click the Browse button next to the "Specify output table" box and store the table as **LandUseSummary** in the \GIST2\MyExercises folder. Click Save. When your dialog box matches the following one, click OK to start the summary process. For more information on performing field summaries, click About Summarizing Data in the summary dialog box.

7 When the process has run, click Yes to add the table to the map document. Close the attribute table and click the List By Source button at the top of the table of contents. You'll see the summary table that you just created.

8 Right-click the table and select Open to view the results.

OID	UseCode	Count_UseCode	Sum_Acreage
0	A1	97	19.1053
1	B1	6	22.4217
2	B4	2	0.3515
3	CRH	3	3.3955
4	F1	61	58.3608
5	VAC	33	48.5091

The table shows many of the results the city manager needed, and just in time for the meeting. Codes are defined as follows:

A1	Single Family Residential
B1	Multi-Family Residential
B4	Duplex
CRH	Church
F1	Commercial
VAC	Vacant

9 Switch to layout view and add a text box to the map that displays the total number of parcels, the total number of dwelling units, the total acreage, and the breakdown of acreage per land-use type. This exhibit will be perfect to distribute at the city council meeting and get the ball rolling on this project.

10 Save your map document as **Tutorial 4-1.mxd** in the \GIST2\MyExercises folder. If you are not continuing on to the exercise, exit ArcMap.

Exercise 4-1

The tutorial showed how to select features inside a box in two ways: manual select and select with a graphic.

In this exercise, you will repeat the process using a different area of the map. There is another area at the north end of town that the city manager has outlined. The manager would like you to calculate the same statistics about that region as you did for the first one.

- Continue using the map document you created in this tutorial or open Tutorial 4-1.mxd from the \GIST2\Maps folder.
- Go to the bookmark Site 2.
- Select the features inside the marked area using one of the techniques from the tutorial.
- Perform statistics or summaries to get the following:
    - Number of parcels
    - Number of dwelling units
    - Total acreage
    - Breakdown of acreage by land use:

        B1 = Multi-Family Dwelling Units
        F2 = Industrial
        POS = Public Open Space (Park)
        VAC = Vacant
- Add these values to your map layout.
- Change elements such as the titles, colors, and legend to make a visually pleasing map.
- Save the results as **Exercise 4-1.mxd** in the \GIST2\MyExercises folder.

WHAT TO TURN IN

If you are working in a classroom setting with an instructor, you may be required to submit the maps you created in tutorial 4-1.

Turn in a printed map or screen capture of the following:

Tutorial 4-1.mxd

Exercise 4-1.mxd

Tutorial 4-1 review

This tutorial showed two techniques for finding what's inside a specified region. The regions were either drawn as a selection box or as a graphic that stays on the map. The first technique worked well, but if you accidentally clear the selected features, you start all over again. The second technique used a graphic element for the selection. The element is more permanent and can be used over and over. It can also be symbolized and used as part of the map layout to mark the area of interest.

Once the features were selected, you used two attribute table functions to get information about the data. The first gave you simple statistics for a selected field. The Statistics command returns the minimum, maximum, sum, count, mean, and the standard deviation. The second command, the Summarize command, lets you do more with a selected field. It returns every unique value in the selected field, then lets you perform any of the statistics operations against the other fields. With this tool, you were able to obtain the sum of the acreage for each land-use type.

4-1

4-2

STUDY QUESTIONS

1. What happens to features that cross the selection boundary?
2. Look at the summary operations for text fields. How do they differ from numeric fields?
3. Is every numeric field suitable for summary operations?

Other real-world examples

The police department might need to close a road for the Independence Day parade. The department could draw a boundary on the map and use it to select property owners who might need to be notified.

After noticing some clogged storm drain inlets, the public works department might draw a boundary around the clogged inlets and use the map to select pipes to inspect.

Tutorial 4-2

Finding what's partially inside

Features that fall completely within the study area boundary are easy to work with, but sometimes the features cross a boundary line. These features must be handled a different way in order to extract correct data from them.

Learning objectives

- *Select features in relation to a boundary*
- *Understand different overlay functions*

Preparation

- *Read pages 87–104 in* The ESRI Guide to GIS Analysis, Volume 1.

Introduction

When using boundaries to select features, not all features will fall completely within the boundary; they may share an edge with the boundary or they may cross the boundary. The attributes associated with the feature represent the whole feature. If only a portion of the feature is inside the boundary, then only a portion of the attribute value can be used for overlay analysis.

As an example, examine some census data. Each census block has a count of how many people are in it. If a feature's boundary crosses through a census block, how many people will be shown to live inside the boundary? Not all of them; only some portion. How much square footage of the census block is inside the polygon? ArcMap calculates area automatically for each polygon feature (Shape_Area), so the census block must be split at the boundary in order to get the total. If the straight totals were used, the results would be wrong.

When the selection of features within a boundary is returning whole polygons or lines, it's OK to get a count or a summary as long as you know you are getting the values for the whole polygons. When features cross the boundary and you need values for just the areas inside the boundary, then you need to split the polygons at the boundary and in some cases proportionally divide the values. For census data, you can calculate the proportional population within the smaller pieces by multiplying the density times the area. It is important to note, however, that the resulting values are a broad estimate, as splitting the values proportionally will not take into account any concentration of values within the polygon. Comparing the results to more accurate data will most certainly highlight the inaccuracies achieved with this method.

Scenario   The city manager took your map to the city council for review. The council member who represents that region remembered that there are some flood zones that would affect this property. The council member would like some of the same calculations performed again, but only for the areas inside the floodplain, including a breakdown of how much of each of the flood zones is in the target area. The city engineer is expecting this information in the morning so that the city can move forward on the project.

Data   The dataset you will be working with is the cadastre layer for Oleander. It represents platted and unplatted property and contains a field for land-use codes (UseCode). The field DU represents the number of dwelling units per parcel, and you'll use that for a housing count. You'll also use the Shape_Area field provided automatically by ArcMap for the area calculations. Overlaid on this are the flood zones as determined by the Flood Insurance Rate Maps (FIRM) from FEMA. You'll use this to determine what's inside the floodplain. It also has the flood zone categories in the field ZONE_:

A	100-year frequency storm (can be filled and reclaimed)
B	50-year frequency storm (can be filled and reclaimed)
FW	Floodway (cannot be reclaimed)
W	Standing water (lakes and ponds)

There's also a field called ZoneName that contains the name of the drainage shed for each flood-prone area.

Select features by location

1   In ArcMap, open Tutorial 4-2.mxd. The map is similar to the previous tutorial, showing the land use and the area of interest. You also see the flood-prone areas. A quick visual inspection shows that some parcels are only partly inside the flood zone, so you will need to do calculations involving just those areas.

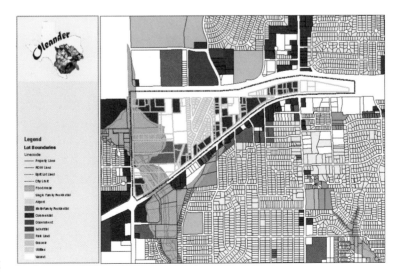

You need to find out how many parcels are in the floodplain, what the total area of land is inside the floodplain, a breakdown of area of land use inside the floodplain, and a breakdown of how much of each flood zone category is in the selected area.

The 202 parcels inside the study area are already selected. If yours become unselected, repeat the procedure from the previous tutorial and use the graphic boundary to select them.

You'll use the flood zone boundaries to select the parcels that intersect them using the Select By Location tool. This tool lets you select features from one layer based on their spatial relationship to features in another layer. First write out what you want to do:

> *I want to select features from the currently selected set of the Land Use layer that intersect the Flood Areas layer.*

Now you will set the parameters in ArcMap to make the desired selection.

2 On the main menu, click Selection > Select By Location.

3 In the Select By Location dialog box, change the Selection method to "select from the currently selected features in."

4 In the Target layers(s) pane, click the box next to Land Use. If your list does not contain all the layers, clear the box next to "Only show selectable layers in this list."

5 Set the Source layer to FloodAreas.

6 Set the Spatial selection method to Target layer(s) features intersect the Source layer feature. There are a lot of options to describe the relationship you want between the two layers. Intersect means that any portion of the parcels falls within the flood area. Some of the others may mean that the features are totally inside the flood area, and some may mean that they only touch the line representing the border of the flood area.

7 When your dialog box matches the graphic, click OK to perform the selection.

ArcMap has made a new selection from the previously selected features showing which parcels intersect the flood area. Notice that some cross the boundary and some are completely within the boundary of the Flood Areas layer.

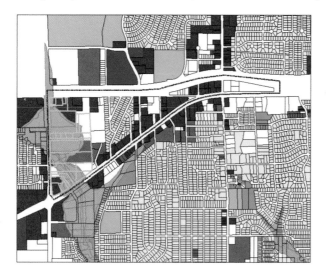

Some of the functions that you want to do will not be affected by the whole parcel being selected. For instance, the first thing you need to know is how many parcels are affected.

8 Click the List By Selection button at the top of the table of contents. How many Land Use parcels are selected?

So your first statistic for the city manager is that there are 23 properties in the study area that are inside the floodplain.

Next you need to know how much acreage that represents. You can't total the acreage, because it would include the areas of some of the parcels that fall outside the floodplain. You will need to perform an overlay operation to cut the parcels at the flood boundary.

Create an overlay using selected features from different layers

1 Open the Search tab and search for the Overlay toolset. Pause your cursor over each of the results to see a description of the tools. The tools that perform the type of overlay you want to do are Intersect, Spatial Join, and Union. If you have an ArcInfo license, you will have more choices in the Overlay toolset. You can read in the ArcGIS Desktop Help what the other tools do, but this tutorial will use Union.

Union will take the input feature classes and merge them together, creating a new output layer with the features split at their computed intersections. The result is that all of the features of both layers are transferred to the output. In addition, all the attributes from both feature classes are preserved in the output.

You will use the Land Use layer and the Flood Areas layer as the input layers.

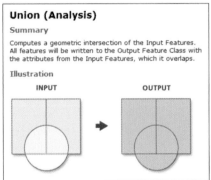

Union (Analysis)

Summary

Computes a geometric intersection of the Input Features. All features will be written to the Output Feature Class with the attributes from the Input Features, which it overlaps.

Illustration

INPUT → OUTPUT

2 In the Search results, click the Union tool to open its dialog box. Use the drop-down box and set Input Features to Flood Areas.

3 Use the drop-down box again to add Land Use to the Input Features list.

4 Click the Browse button next to "Output Feature Class," name this feature class **FloodAreaOfStudy** (do not include any spaces), and store it in LandUseReport.mdb in the \GIST2\MyExercises folder.

5 When your dialog box matches the graphic to the right, click OK.

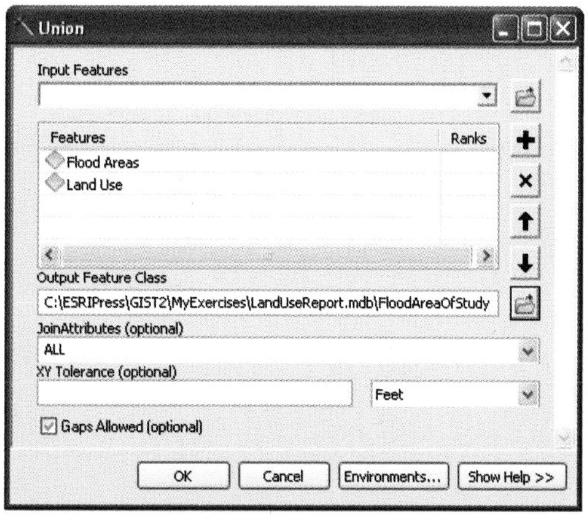

The resulting feature class has the selected parcels that fell within the floodplain, as well as the floodplain from throughout the city. Some portions of the parcels that are not within the floodplain are still in this layer, and some portions of the floodplain that are not part of this analysis are still in the output layer. You can remove them with a definition query. ArcMap makes this easy to do. Any parcel feature that did not intersect the flood zone layer will have an FID_FloodZone value of -1, and any floodplain area that did not intersect a parcel will have an FID_Parcels value of -1. The definition query will exclude features with an FID_FloodZone of -1 and an FID_Parcels of -1. The query includes the operator "< >", which means "not equal to."

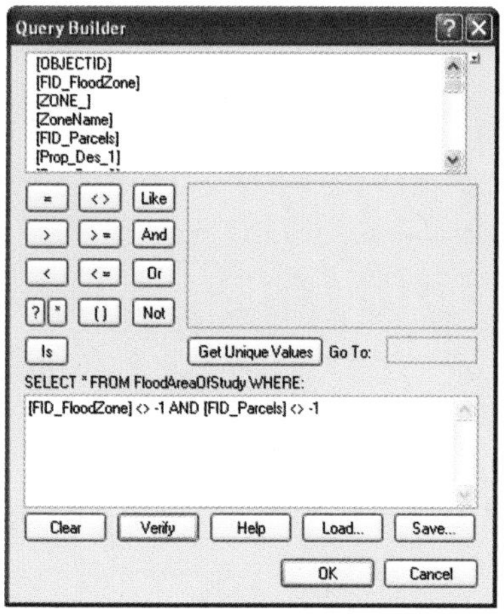

6 Switch back to List By Drawing Order in the table of contents and right-click the FloodAreaOfStudy layer, then click Properties. Go to the Definition Query tab and select Query Builder. Build the query shown in the graphic and verify it. Then click OK and OK again. The FloodAreaOfStudy layer now shows only the areas that are both within the selected set of parcels and the floodplain.

Next you want to calculate acreage. There's a field in the attribute table called Acreage that you may be able to use.

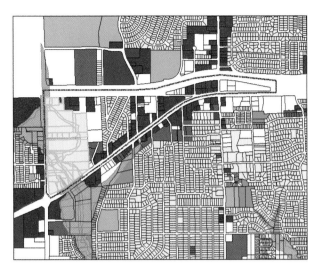

Calculate acreage

1 Right-click the FloodAreaOfStudy layer and open the attribute table. Look for the Acreage field.

There are values in the Acreage column, but some values repeat. This is because ArcMap didn't automatically update the areas represented here; it only updated the field Shape_Area. You need to recalculate the acreage value for this field.

2 Right-click the field name Acreage and select Calculate Geometry. Click Yes on the warning. In the Calculate Geometry dialog box, select Acres US [ac] in the Units drop-down list and click OK. You are finally ready to do the area calculations for the Land Use types and the Flood Zone categories. The process will involve doing a field summary on both the UseCode field and the ZONE_ field.

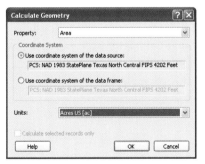

3 Right-click the field UseCode and select Summarize. In the dialog box, expand the choices under Acreage and check the box next to Sum.

4 Click the Browse button next to "Specify output table," name the table **StudyAreaUse**, and store it in the \GIST2\MyExercises folder. Click Save. When your dialog box matches the graphic, click OK. When prompted, select Yes to add the table to the current map document. The results are recorded in a new table that includes the total acreage for each unique land-use code.

[Summarize dialog box showing: UseCode selected as field to summarize; summary statistics list with FID_Parcels, Prop_Des_1, Prop_Des_2, Acreage (with Minimum, Maximum, Average, Sum checked, Standard Deviation, Variance); output table path C:\ESRIPress\GIST2\MyExercises\StudyAreaUse]

4-1
4-2

YOUR TURN

The city manager also wants a similar summary operation to show acreage of each of the flood zone categories.

Repeat the summarize process using the ZONE_ field to get the total acreage for each flood zone area. Name the output file **StudyAreaZone** and add it to the map document. When you're done, close the attribute table for the FloodAreaOfStudy layer.

5 On the List By Source pane of the table of contents, right-click StudyAreaUse to open the table. Here's the next set of data that the city manager needs. The acreage of each use-code value has been summed in this table.

OID	UseCode	Count_UseCode	Sum_Acreage
0	B1	24	11.8586
1	F1	25	14.1502
2	VAC	34	11.3089

6 Right-click the StudyAreaZone file and open it. This table has the acreage for each flood zone category. Totaling the Sum_Acreage field will yield the total acreage inside the flood area—the final information you needed to find.

OID	ZONE_	Count_ZONE_	Sum_Acreage
0	A	37	26.3248
1	B	19	7.2477
2	FW	27	3.7432

7 When you have finished looking at the tables, close them. Add another text box to your layout and add all the information that the city manager wants.

8 Save your map document as **Tutorial 4-2.mxd** in the \GIST2\MyExercises folder. If you are not continuing on to the exercise, exit ArcMap.

Exercise 4-2

The tutorial showed how to select features using features of another layer. In this exercise, you will repeat the process using a different area of the map. There is another area at the north end of town that the city manager has outlined. The manager would like you to calculate the same statistics about that region as you did the first one.

- Continue using the map document you created in this tutorial or open Tutorial 4-2.mxd from the \GIST2\Maps folder.
- Go to the bookmark Site 2.
- Select the features inside both the study area and the flood zone using the techniques from the tutorial.
- Perform statistics or summaries to get the following:
  - Number of parcels
  - Total acreage
  - Breakdown of acreage inside the flood zone by land use
  - Breakdown of acreage inside the flood zone by flood-zone category
- Add these values to your map layout.
- Change elements such as the titles, colors, and legend to make a visually pleasing map.
- Save the results as **Exercise 4-2.mxd** in the \GIST2\MyExercises folder.

WHAT TO TURN IN

If you are working in a classroom setting with an instructor, you may be required to submit the maps you created in tutorial 4-2.

Turn in a printed map or screen capture of the following:

Tutorial 4-2.mxd

Exercise 4-2.mxd

Tutorial 4-2 review

Not all boundaries used for analysis are going to be graphics or user-drawn boundaries. Many times they will be existing features. Using the Select By Location tool, you were able to use one set of features to select another set that had a spatial relationship. The relationship could have been defined as inside the boundary, completely inside the boundary, crossing the boundary, or other options.

Merely selecting the features enabled you to do certain attribute calculations. You were able to get a count of features inside the boundary, and make a list of the features. But because some of the features crossed the boundary, you were not able to get a total area. In order to do this, you had to split the polygons at the boundary to create new features, a process described in *The ESRI Guide to GIS Analysis, Volume, 1* as feature overlay analysis.

4-1
4-2

STUDY QUESTIONS

1. What must be done to data summarized by area when its features are split at a boundary?
2. Can any feature type be used in a Select By Location operation?
3. Can any feature type be used with the feature overlay analysis tools?

Other real-world examples

A building department might make a map of all the inspections that are scheduled for today. Then it could use the predefined inspection district boundaries to assign the inspections to the various inspectors.

A county voter registration office might select the number of households that fall within each voting precinct. This would help it determine how many poll workers each site might need.

School districts might map attendance boundaries for each of the elementary, junior high, and senior high schools. Then these boundaries might be used to determine how many students will be attending each school, broken down by grade level, and the appropriate number of textbooks could be ordered.

5

Finding what's nearby

Not every important geographic relationship is based on overlapping or adjacent features. Many are based on the idea that they are nearby or within a distance of each other. Near, however, is rather subjective; what is near to some may not be to others. So a near value should be associated with the type of data and the situation in which the data exists.

Tutorial 5-1

Selecting what's nearby

One of the most powerful types of spatial analysis is finding features within a specified distance of other features. This can be a straight-line distance or distance along a network or path. Selected features can be displayed for visual analysis or put into separate feature classes for use in other analyses.

Learning objectives

- *Select features*
- *Analyze distance relationships between features*

Preparation

- *Read pages 115–128 in* The ESRI Guide to GIS Analysis, Volume 1.

Introduction

In the previous tutorial, you used features to select other features. In this tutorial, you'll use features and a distance value to find other features within that distance.

You'll also discover that the process doesn't have to be very complicated, if all you need is a list or count of the nearby features.

ArcMap has a selection tool called Select By Location, which you used in the previous tutorial. A straight-line search distance can be added to the parameters of this tool to give it the added power to select things nearby.

This selection tool has two downsides. The distance is "as the crow flies," so it does not take into account street accessibility, terrain, or physical obstacles. Imagine if the analysis showed some great property within your search distance, but it's on the other side of a lake that would take five times the travel distance to get there.

The other drawback to the selection process is that the selection boundary is not drawn as a feature. The selected features are shown, but the boundary is not.

If the data does not have constraints such as those listed above, and the selection boundary does not need to be shown, then this can be a quick and powerful analysis tool.

Scenario   The Tarrant County Health Department wants to come out to Oleander to do some mosquito control work. Workers will be walking the creeks looking for standing water, debris that may be clogging the creek, and harborages for mosquito larvae. Before the workers arrive, however, the department would like to mail out notices to all the property owners within 50 feet of the creek and let them know when the work will be done.

Data   The map document includes two layers. The first is the parcel dataset for the City of Oleander. Use the fields Addno, Prefix, StName, Suffix, and SufDir to get the street address for each selected parcel.

The other dataset is the creek data from the North Central Texas Council of Governments, which maintains a lot of regional data.

Create a visual buffer

1 In ArcMap, open Tutorial 5-1.mxd. The map shows the first area that the county wants to clean up. The purple line is the creek. You'll use it as the selected feature and find parcels that are near it.

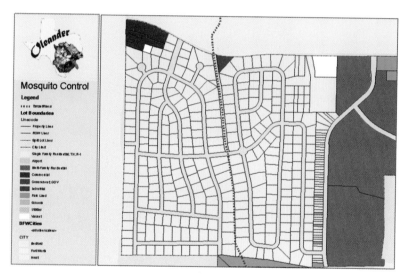

There's a tool that will give a great visual of the target features. Try that first and see how it works. First, you'll have to customize the standard ArcMap toolbar to include this tool.

2 On the main menu, select Customize > Customize Mode.

5-1
5-2
5-3
5-4
5-5
5-6
5-7
5-8
5-9

3 Go to the Commands tab, scroll through the Categories, and click Selection. In the right window, click the Buffer Selection tool and drag it to the Tools toolbar. When you have it on the toolbar, release the mouse button. Close the Customize dialog box.

While dragging a tool to a toolbar, a black bar will be displayed to show where the tool will be placed when released. When you have time, scroll through these categories of tools in the Customize dialog box. These tools are not on the default toolbars but add great functionality to the program. Some of the commands can be found in menus, so if you use the commands frequently it's helpful to add a button on the Tools toolbar. A search box on the Commands tab can help you find tools, or you may just browse by category.

4 Click the new Buffer Selection tool [icon]. Select the TarrantRivers layer and set the distance to 50 feet. Then click OK.

5 Using the Select Features tool [icon] ▾, click the purple creek feature to select it. The Buffer Selection tool draws a 50-foot buffer around the selected features. Look closely at the creek and you'll see the buffer.

This is great for visual analysis. When you print the document, the selection buffer will print. This will let you clearly see which parcels will get notified.

Unfortunately, it is a visual reference tool only. No features in the parcels dataset were selected, so there's no way to make a list. This tool is better suited for other purposes, so you'd better try another method.

6 Open the Buffer Selection tool again, select TarrantRivers, and set the buffer distance to **0**. Click OK to close the dialog box. Previously you used the Select By Location tool. One parameter of this tool that you did not use is the ability to select within a distance of the selected features. The creek feature should still be selected from the last procedure, but if not, just select it again.

Select by location with a distance buffer

1 On the main menu toolbar, click Selection > Select By Location to open the dialog box.

You need to write out the objective of the search so that you can understand what parameters to use in the selection dialog box:

> *I want to select features from the target layer Land Use that are within 50 feet of the features in the source layer TarrantRivers.*

2 Fill out the selection dialog box as shown in the image. Make sure that both the "Use selected features" and "Apply a search distance" boxes are checked. Set the distance to **50** feet. When your dialog box matches the image, click OK to make the selection. You should have a selection, but the cyan outlines won't look very good on the printed map. You will change the selection color to red, but only for the Land Use layer.

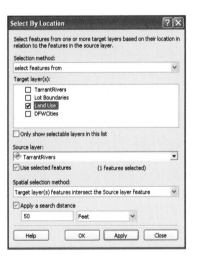

Notice that even though you used a buffer distance in the selection, there is no buffer ring drawn on the map.

5-1
5-2
5-3
5-4
5-5
5-6
5-7
5-8
5-9

Set the selection symbol and list the results

1 Right-click the Land Use layer and open the properties. Go to the Selection tab. Check the radio button next to "with this symbol." Then click the cyan box to open the Symbol Selector dialog box. In the Symbol Selector, set the outline color to Flame Red and the outline width to 4. Click OK, then OK again to close the properties.

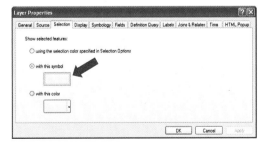

The map now shows clearly which parcels fall within the selection zone of the creek. Since you are only after a list of addresses, you don't need these features to be in a separate feature class. You only need to get the list from the attribute table.

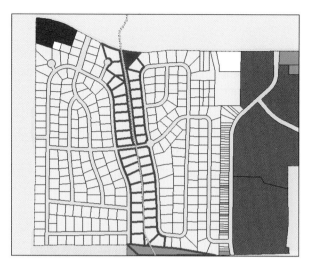

2 Right-click the Land Use layer and select Open Attribute Table. Click the Show selected records button at the bottom of the table to view only the selected records. There's the list, which can be exported to a DBF table and used to create mailing labels.

3 If you are working on this tutorial in a classroom environment, print the list for the completed project (Table Options > Print). When you are done, close the attribute table, and click Selection > Clear Selected Features. Notice what happens to the map.

4 Save your map document as **Tutorial 5-1.mxd** in the \GIST2\ MyExercises folder. If you are not continuing on to the exercise, exit ArcMap.

Exercise 5-1

The tutorial showed how to use a selected feature, and the Select By Location tool, to select land-use parcels within 50 feet of the selected creek.

In this exercise, you will repeat the process for one of the other creeks the county wishes to inspect.

- Continue using the map document you created in this tutorial or open Tutorial 5-1.mxd from the \GIST2\Maps folder.
- Go to the bookmark Creek Area 2.
- Select the creek in that region.
- Use the Select By Location tool to select all the land-use parcels within 50 feet of the creek.
- Print the resulting list of addresses.
- Change elements such as the titles, colors, and legend to make a visually pleasing map.
- Save the results as **Exercise 5-1.mxd** in the \GIST2\MyExercises folder.

WHAT TO TURN IN

If you are working in a classroom setting with an instructor, you may be required to submit the maps you created in tutorial 5-1.

Turn in a printed map or screen capture of the following:

Tutorial 5-1.mxd and a printed list of the selected addresses

Exercise 5-1.mxd and a printed list of the selected addresses

5-1
5-2
5-3
5-4
5-5
5-6
5-7
5-8
5-9

Tutorial 5-1 review

The selection processes in this tutorial used a straight-line distance from the selected features to select other features. A buffer zone is temporarily created by the software, and features that touch that zone are selected. The buffer is never really seen and can't be drawn on a map; it existed only for the moment necessary to make the selection.

The selected set can be used to get lists or counts of features within the distance. But since no hard boundaries are drawn, that's the limit of what can be done with the selection. If the desired result is to measure areas or calculate densities, then other more advanced tools would need to be used.

STUDY QUESTION

1. When would it be appropriate to do summary operations against the selected data?

Other real-world examples

The public works department might use a straight-line distance selection to get a list of property owners along a street that is being scheduled for repairs.

The health department might identify a hazardous site such as an abandoned lead smelter. They would select houses within a certain distance of the site to alert residents of upcoming cleanup efforts.

The economic development department might select census tracts around a business center to compile demographic data about the area.

Tutorial 5-2

Creating buffer features

A straight-line distance buffer can be created around selected features. This new feature can be symbolized on the map and used in overlay analysis to clip other features.

Learning objectives

- *Select features*
- *Create buffers*

Preparation

- *Review pages 115–128 in* The ESRI Guide to GIS Analysis, Volume 1.

Introduction

Buffers can be created on a more permanent basis than the selections you did in the previous tutorial. They can be saved as their own feature class, letting you then do the same type of inside–outside analysis as before. The difference now is that the region we're using was created by applying a distance to a selected set of features.

Again, you have the choice of three overlay analysis tools to use with the buffers: Identity, Intersect, and Union. Review these in the ArcGIS Desktop Help to determine their functionality.

Once you do the overlay function, the resulting feature class can be used for calculations such as sum of area. This also creates a feature that can be symbolized in the map showing the area of study.

Scenario   Every time a zoning change is proposed, the property owners within 200 feet of the subject tract must be notified by mail. They are given the date and time of the city council meeting at which the change will be discussed. You've been told of a certain tract of land where a zoning change is being proposed, and the city secretary is requesting a list of adjacent property addresses.

The process is pretty simple. You'll select the subject tract, buffer it 200 feet, and select the parcels that intersect the buffer. You could do this with only a selection, but the map needs to have a graphic showing the notification boundary.

Data All you need for this tutorial is the parcel dataset for the City of Oleander. You'll be using the fields Addno, Prefix, StName, Suffix, and SufDir to get the street address for each selected parcel.

Create a buffer

1 In ArcMap, open Tutorial 5-2.mxd. The purple-hatched parcel is the one subject to a zoning change. The property owner wants to change from C-2 (Community Business District) to C-2(A) (Community Business District with Alcohol Sales) for a new convenience store.

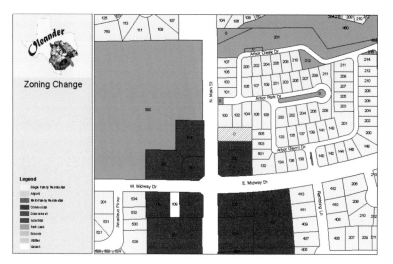

The first thing you'll need to do is create a 200-foot buffer around the subject tract. It is already selected, and the selection color is the purple hatch that you see.

2 Click the Search tab and enter **buffer**. Click Buffer to open the buffer dialog box.

3 In the Buffer dialog box, click the drop-down menu for Input Features and select the Land Use layer.

4 Name the output feature **ZoningCase1** and store it in the MyData.mdb geodatabase in the \GIST2\MyExercises folder.

Output Feature Class

Look in: MyData.mdb

Name: ZoningCase1   Save

Save as type: Feature classes   Cancel

5 Set the linear distance to **200** and the units to feet. When your dialog box matches the image, click OK.

Buffer

Input Features
Land Use

Output Feature Class
C:\ESRIPress\GIST2\MyExercises\MyData.mdb\ZoneCase1

Distance [value or field]
● Linear unit
                                                    200   Feet
○ Field

Side Type (optional)
FULL
End Type (optional)
ROUND
Dissolve Type (optional)
NONE
Dissolve Field(s) (optional)
☐ UseCode
☐ Prefix
☐ StName
☐ Suffix

OK   Cancel   Environments...   Show Help >>

A buffer feature is created around the selected parcel. Notice that it's not a circle! The buffer tool buffers the edge of the parcel polygon, so the resulting feature takes on the shape of the original feature. This is the desired result, but it can be made a little prettier for your map.

5-1
5-2
5-3
5-4
5-5
5-6
5-7
5-8
5-9

Update the symbology

1 Right-click the ZoningCase1 layer and open the properties. Go to the Symbology tab.

2 Click the Symbol box to open the Symbol Selector, and then select Lilac. Click OK.

3 Go to the Display tab. Set the Transparent value to 60%. Click OK to close the Properties dialog box. That looks a lot better. The map clearly delineates the subject tract and the 200-foot boundary.

4 Print or save your map document as **Tutorial 5-2a.mxd** in the \GIST2\ MyExercises folder before you move on to the selection step, since the selection will turn a lot more parcels purple.

The next step will be to do the selection process. Once again, you'll use the Select By Location tool and build a selection sentence. You don't need to specifically select the buffer feature because it is the only feature in that feature class. You also don't need to use a distance with the selection since the buffer already represents the 200-foot buffer that you need.

The selection sentence will be the following:

> *I want to select the features from the Land Use layer that intersect the features in the ZoningCase1 layer.*

View the attributes of selected features

1 On the main menu, click Selection > Select By Location. Determine on your own what the selection parameters will be; then start the Select By Location tool. When you have confirmed that your dialog box matches the following image, click OK. The selected features turn purple because you set the layer properties to do this. All you need is the list from the attribute table.

2 Right-click the Land Use layer and select Open Attribute Table. Click the Show selected records button to view only the selected features.

5-1
5-2
5-3
5-4
5-5
5-6
5-7
5-8
5-9

Customizing a menu

Notice the extra tool on the menu in the previous image: Open Table Showing Selected Features.

This tool saves the step of clicking the Selected button after a table is opened. The tool can be added (if you don't already have it) in the same manner that you added the Buffer Selection tool before.

1. On the main menu, select Customize > Customize Mode.

2. Click the Commands tab on the Customize dialog box and scroll to the Layer category.

3. On the Toolbars tab, check the box next to Context Menus. In the resulting pop-up menu, click the drop-down arrow and choose Feature Layer Context Menu.

4. Find Open Table Showing Selected Features in the Commands list and drag it to the Feature Layer context menu.

5. Close the Customize Mode dialog box. Right-click a layer to test the new command.

The resulting table can now be used to create mailing labels, printed to show a list of properties, or exported to a table and saved for future use.

Exporting a table

The results of this analysis could be very useful in a spreadsheet or used to perform a mail merge using a word-processing software. To export just the tabular data of a layer, open the layer's attribute table and click the Table Options button. There you will find the Export tool. The data can be exported as a personal or file geodatabase table, as a dBase file, or as a delimited text file. Make sure to notice the Export drop-down menu to export all records or just the currently selected features.

3   Print or save your map document as **Tutorial 5-2b.mxd** in the \GIST2\MyExercises folder. If you are not continuing to the exercise, close ArcMap.

Exercise 5-2

The tutorial showed how to create a buffer feature from the selected features. A distance was specified, and a new feature was created by extending the boundary of the selected feature by the given distance.

In this exercise, you will repeat the process using a different parcel. The property owner requesting the zoning change in the tutorial has selected a second lot for another location. The request also is to change from C-2 (Community Business District) to C-2(A) (Community Business District with Alcohol Sales).

- Continue using the map document you created in this tutorial or open Tutorial 5-2.mxd from the \GIST2\Maps folder.
- Move to the bookmark Zoning Case 2.
- Select the vacant lot on the southwest corner of North Main Street and Bear Creek Drive.
- Create a 200-foot buffer feature around the subject tract.
- Select the parcels that will need to be notified of the impending zoning change.
- Save the results as **Exercise 5-2.mxd** in the \GIST2\MyExercises folder.

5-1
5-2
5-3
5-4
5-5
5-6
5-7
5-8
5-9

WHAT TO TURN IN

If you are working in a classroom setting with an instructor, you may be required to submit the maps you created in tutorial 5-2.

Turn in a printed map or screen capture of the following:

Tutorial 5-2a.mxd, Tutorial 5-2b.mxd, and a printout of the selected parcel addresses

Exercise 5-2.mxd and a printout of the selected parcel addresses

Tutorial 5-2 review

Unlike a straight-line distance selection, this time you created a new feature to use for the selections. The buffer that was created is the result of a straight-line measurement from the selected feature, whether it is a point, line, or polygon.

With the buffer being a separate feature, things can be done with it that could not be done from the straight-line selection process. Most importantly, it can be symbolized on the map. Also, since it is a feature, its area can be measured.

Another important difference would be seen if the buffer was done on multiple features. The buffers can each produce a unique selection set, while the straight-line selection will always produce a single set of features as a result.

STUDY QUESTIONS

1. Will the selection with a buffer always yield the same results as the straight-line selection?
2. Name three characteristics of a buffer feature that a straight-line selection does not have.

Other real-world examples

The police department buffers schools, playgrounds, and arcades to produce zones of strict drug enforcement. Each of these types of facilities is given a different buffer size, which could not be done with a straight-line selection. The buffers are also depicted on a map to show field officers the boundaries of the zones.

The local hospital buffers its facilities 500 feet to produce a quiet zone. A printed graphic of the zones is distributed to local citizens. It may also be used by the police department to help enforce the restrictions.

Tutorial 5-3

Clipping features

Besides providing a symbolized graphic on the map, a buffer layer can be used with the overlay analysis tools. This will cut features out of other layers, allowing you to determine the area of features inside the buffer.

Learning objectives

- *Select features*
- *Create buffers*
- *Analyze overlays*

Preparation

- *Review pages 115–128 in* The ESRI Guide to GIS Analysis, Volume 1.

Introduction

The feature created with the Buffer tool is just that, a feature. Just like any other polygon, it can be used with the overlay analysis tools. Features can be clipped using the buffer ring like a cookie cutter. The resulting features are clipped at the boundary and have a new area.

With the selection tools, you could only get lists and counts of features. Once the features are clipped, you can do summaries on area, and with density values you can calculate the proportion or percentage of a value within the buffer.

Scenario The zoning notification process from the previous tutorial has been challenged by the homeowners group adjacent to the subject tract. All of the owners of the residential lots signed the petition to block the zoning change. The city council still votes and decides the issue. However, if 60 percent of the area inside your notification buffer is residential, the city council vote would have to be a super majority (7 to 2) in order for the ordinance to pass.

You will need to provide the city attorney an exhibit showing the percentage of each land-use type inside the notification zone. He will advise the city council what the vote count would need to be in order for the requested zoning change to be approved.

Data All you need for this tutorial is the parcel dataset for the City of Oleander. You'll be using the field UseCode and Shape_Area to get the data you need.

Dissolve a layer

1 In ArcMap, open Tutorial 5-3.mxd. Notice the buffer feature you created in the previous tutorial. You will be using the buffer feature as a "cookie cutter" to clip out the parcels that fall within it.

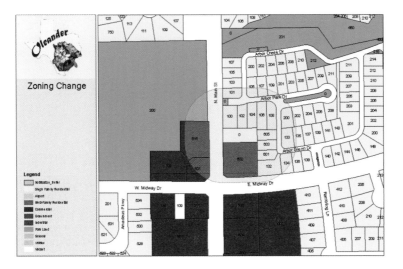

This time you can't just use the Select Features tool because parcels that are partially inside the buffer would be selected in their entirety. Instead you'll use the Intersect tool to cut out the desired parcels. The resulting parcel segments can be used to get an exact area inside the buffer.

When the Intersect tool is run, it will act upon only the selected features—unless no features are selected—in which case it will act upon the entire dataset. In some cases, this can be very useful. Imagine having a countywide dataset from which you wanted to do an intersect in only a certain region. If you selected that region first, the unselected features would not be used in the intersect process, thus saving a lot of processing time. In this case, it is OK to have the Intersect tool use the entire Parcels dataset. If you wanted to check for selected features, you could go to the Selection tab at the bottom of the table of contents and see which layers have selected features. In this particular analysis, it is best to start with no selected features.

2 On the Tools toolbar, click the Clear Selected Features button if necessary to ensure that no features are selected.

3 Open the Search tab and type in **Intersect**. Click the tool in the results box to run it. The Intersect tool will output a new feature class that contains the areas common to both input layers.

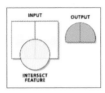

The Intersect tool computes a geometric intersection of the Input Features. Features or portions of features that overlap in all layers or feature classes will be written to the Output Feature Class.

4 In the Input Features drop-down menu, select Land Use to add it to the list of layers.

5 Repeat the process to add the Notification_Buffer layer to the list.

6 Name the output file **ZoneCase1LandUse** in the MyData.mdb geodatabase in the \GIST2\MyExercises folder. When your dialog box matches the graphic, click OK to run the Intersect process.

A new layer is created and added to the map document. This layer is the clipped features common to both input layers.

You need to get the percentage of each land-use type in this new coverage. To do this, you'll dissolve the layer by the UseCode field, creating a new feature class with one polygon for each unique land-use code. Each will have an area calculated for it, and you can use that to get the percentage.

7 Use the Search dialog box to find and run the Dissolve tool.

8 In the Dissolve dialog box, set Input Features to ZoneCase1LandUse.

5-1
5-2
5-3
5-4
5-5
5-6
5-7
5-8
5-9

9 Save the Output Feature Class as **LandUseDissolve** in the MyData geodatabase in the \GIST2\MyExercises folder.

10 Set the Dissolve Field to UseCode by checking the box next to it in the list. Note that the UseCode field was duplicated in the intersect process because it existed in both input feature classes. When your dialog box matches the graphic, click OK to start the dissolve process.

Analyze the dissolved layer's data

The dissolve process creates a new feature class, but in reality you only need the data it contains. The attribute table for this new layer contains the area for each unique occurrence of UseCode in the original file. Take a look at the table, then add a field to calculate the percent of the total. Remember that if the residential category exceeds 60 percent, the city attorney will require a super-majority vote at the city council meeting.

1 Right-click the LandUseDissolve layer and select Open Attribute Table. Note the UseCode field values and the ArcMap-generated Shape_Area field. Right-click the Shape_Area field and select Statistics. Make a note of the Sum value. Be aware that yours may be slightly different from the value shown in the following image. Close the Statistics window.

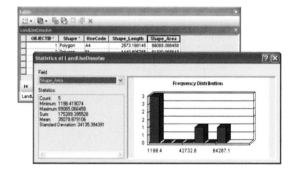

2 Click the Table Options button at the top of the attribute table and select Add Field. In the Add Field dialog box, type in **Percentage** for Name. Click the Type drop-down menu and select Float. Click OK to create the new field.

3 In the attribute table, right-click the new Percentage field and select Field Calculator. Click Yes if you receive a message about calculating outside of an edit session.

4 In the Field Calculator dialog box, double-click the Shape_Area field to add it to the calculation. Type the rest of the calculation as **/175399.395528 * 100.** Click OK.

5 Click OK to calculate the new field values.

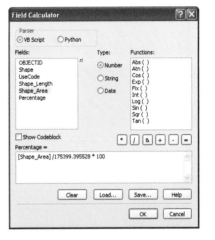

Change a table's appearance

You can see that the percentage of residential property (A4) does not exceed 60 percent. Your value may differ from the value in the image to the right. This data can be added to the map layout, but first you need to clean it up and set some symbology so that it will look nice.

1 At the top of the table, click the Table Options button and select Appearance.

2 Set the table font to Arial Black and the size to 10 and then click OK to finish. This will make the font easy to read on the final map, but you don't need all the extra fields to be visible. You'll turn some of them off and change their names to aliases that make more sense. You'll also format the numbers for easier readability.

3 Move the table out of the main map window or dock it to the bottom of the display. Open the properties of the LandUseDissolve layer and go to the Fields tab. Clear all the field check boxes except UseCode and Percentage.

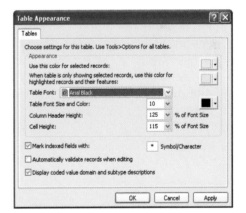

5-1
5-2
5-3
5-4
5-5
5-6
5-7
5-8
5-9

4 Highlight the UseCode field. In the Appearance pane, change the Alias to **Land Use Code**. Then highlight the Percentage field and make its Alias **% of Total Area**.

5 Highlight the Number Format line and click the Number Format button. In the resulting window, set the number of decimal places to 1.

6 Click OK, then OK again. Resize the field widths and table box so that just the data is visible.

7 Click the Table Options button and select Add Table to Layout. Then close the table.

8 The table is dropped into the middle of the map. Drag it over to the left and place it on the title bar. Make sure that the only visible layers are Land Use, Notification_Buffer, and Lot Boundaries. Much of the work you did on this data, and some of the layers you created, are not visible on the final map. They were important steps in calculating the desired result: the percentage of land use inside the buffer. They didn't, however, add much graphically to the map layout.

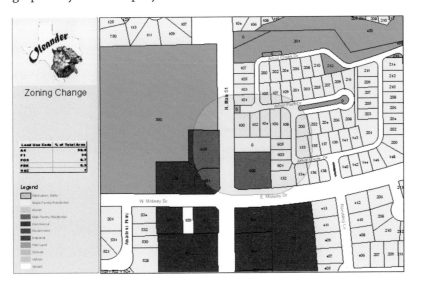

The percent of residential land use in the area did not require the city council to achieve a super-majority vote to either pass or deny this request. Oddly enough, the city council unanimously voted down this request because of what else was nearby as demonstrated in your exhibit: two city parks. The land remains vacant.

9 Save your map document as **Tutorial 5-3.mxd** in the \GIST2\MyExercises folder. If you are not continuing on to the exercise, exit ArcMap.

5-1
5-2
5-3
5-4
5-5
5-6
5-7
5-8
5-9

Exercise 5-3

The tutorial showed how to clip a layer using a buffer as the boundary. The resulting table was used for a generalization operation to extract the necessary data.

In this exercise, you will repeat the process using a different parcel. With the original project denied, the applicant is putting all efforts into getting a second-choice site approved. The homeowners in that area have also filed a petition to force a super-majority vote, and the city attorney has asked for the same percentage breakdown of land-use types.

- Continue using the map document you created in this tutorial or open Tutorial 5-3.mxd from the \GIST2\Maps folder.
- Move to the bookmark Zoning Case 2.
- You will be working with the vacant lot on the southwest corner of North Main Street and Bear Creek Drive.
- Use the buffer created in exercise 5-2 along with the Intersect and Dissolve tools to derive the necessary information.
- Create a summary table to include on the layout.
- Save the results as **Exercise 5-3.mxd** in the \GIST2\MyExercises folder.

WHAT TO TURN IN

If you are working in a classroom setting with an instructor, you may be required to submit the maps you created in tutorial 5-3.

Turn in a printed map or screen capture of the following:

Tutorial 5-3.mxd

Exercise 5-3.mxd

Tutorial 5-3 review

Buffers used for selections could only provide lists and counts, but with the overlay tools it is possible to start calculating areas of intersection. The same type of buffer is created as the selections, but now the overlay tools can be used to clip out only the portions of features that fall within the buffer. Once these are placed into a new feature class, they can be manipulated separately from the original features.

It is important to note that any summary values that might be true for the whole piece are now no longer representative of the area. If a census block with a value of 200 people is split in two, each half will still retain the value of 200. Summary statistics need to be calculated as densities before the overlay operation to retain their relationship to the new area.

STUDY QUESTIONS

1. What is the difference between Intersect and Identity?
2. When is Union the best choice?

5-1
5-2
5-3
5-4
5-5
5-6
5-7
5-8
5-9

Other real-world examples

The health department might buffer water wells to show a spill hazard area and use the buffer to clip out the soil polygons. The clipped features could be used to get a total amount of different soil types in the hazard area in order to evaluate soil permeability.

The regional water authority may buffer a lakeshore by the distance of an access easement and use the buffer to intersect property records. This will tell the percent of the easement that is owned by each property owner.

Tutorial 5-4

Buffering values

Buffering can be done with a single straight-line distance for all features. Or, with an optional setting, each feature can be buffered by a separate value. These values are stored in an attribute field and can be referenced in the Buffer command.

Learning objectives

- *Work with buffers*
- *Work with attribute values*
- *Perform distance analysis*

Preparation

- *Review page 124 in* The ESRI Guide to GIS Analysis, Volume 1.

Introduction

The single distance buffer lets you define one distance for all the features. In the Buffer command, there is an option to derive the buffer distance from a field in the attribute table. This means that every feature could have a unique buffer distance.

When the buffers are drawn, the option of merging the output features becomes critical. Not merging the buffers will create a separate polygon for each feature in the input file. This may result in a large number of polygons. The option to merge them will dissolve the boundaries where the buffers overlap and make fewer polygons with larger areas.

The decision to merge may depend on the ultimate use of the buffers. For visual analysis, the dissolve method looks best. But if a one-to-one relationship between the buffers and the origin features needs to be preserved, then the dissolve function should not be used.

Imagine a stream network with continuous and intermittent streams, each needing a buffer of a different distance. The required distance could be saved in a field and used in the Buffer tool. Where two streams come together, the buffers could be made to merge together, with one buffer being derived from several stream segments. If it were important to preserve the names of the streams in the output buffer features, the dissolve would not be used and a one-to-one relationship would be maintained.

Scenario   The city council has received a report on the effects of noise pollution caused by cars driving on city streets. In an effort to get an idea of how much of the city might be affected by street noise, the council has asked you to prepare an exhibit map. A consultant has sent the noise coefficients for each street in the city, and they have been put into a field in the attribute table. You need to buffer the street centerlines to visually show the area affected by vehicle noise.

Data The street centerline data contains a field called CO_Code that represents the distance that noise of a high decibel level travels from each street.

The parcel layer is included for background interest.

Search for a tool

1 In ArcMap, open Tutorial 5-4.mxd. Here's a pretty plain looking map. The street centerline file is included in the table of contents, but it is not visible. You really don't need to see it; you'll just use it in a buffer analysis.

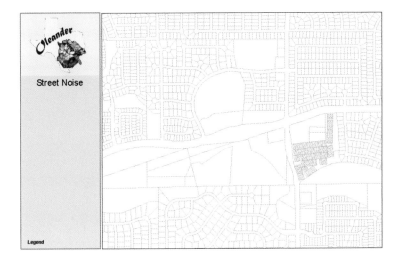

2 Open the Search dialog box and type in **buffer**. Click the Buffer tool in the results window.

Set buffer parameters

1 In the Buffer dialog box, set Input Features to StreetCenterlines.

2 Set the Output Feature Class to \GIST2\MyExercises\MyData.mdb and name the feature class **NoiseBuffers**.

3 Under Distance, click the radio button next to Field. In the drop-down box, set the field to CO_CODE.

5-1
5-2
5-3
5-4
5-5
5-6
5-7
5-8
5-9

4 Set Dissolve Type to ALL. When your dialog box matches the graphic, click OK.

As expected, you got a wider buffer on streets that are creating more noise and a smaller buffer on streets that are generating less noise. The city council will now be able to visualize the table of values that the consultant delivered.

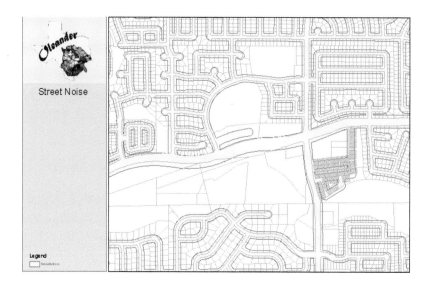

5 Save your map document as **Tutorial 5-4.mxd** in the \GIST2\MyExercises folder. If you are not continuing on to the exercise, exit ArcMap.

Exercise 5-4

The tutorial showed you how to create a multidistance buffer using a value from a field in the attribute table.

The city council reviewed your map, but it would like to see the relationship between speed limit and the measured noise values. In this exercise, you will create a set of multisized buffers using the speed limit field. Overlay the results on top of the noise buffers for a visual comparison.

- Continue using the map document you created in this tutorial or open Tutorial 5-4.mxd from the \GIST2\Maps folder.
- Use the Buffer tool and the field SpeedLimit to create the buffers.
- Symbolize and arrange the layers to make the best map possible.
- Save the results as **Exercise 5-4.mxd** in the \GIST2\MyExercises folder.

WHAT TO TURN IN

If you are working in a classroom setting with an instructor, you may be required to submit the maps you created in tutorial 5-4.

Turn in a printed map or screen capture of the following:

Tutorial 5-4.mxd

Exercise 5-4.mxd

Tutorial 5-4 review

Although this process also creates a straight-line buffer, it's unique in its ability to vary the size of the buffer. Some data preparation must be done to create a field and populate it with a buffer distance.

The field chosen for the buffers is required to be a quantity in map units. Quantities that do not represent a measurable distance such as traffic counts or speed limits would not be suitable. Codes that represent a quality of the road would also not be suitable, since they do not represent a distance. It is also important to note that the field must be represented in map units, since the Buffer tool does not allow unit conversions. So if the map units are meters, be careful not to use values measured in feet for the buffers.

The different size buffers can also be created as single pieces as opposed to merging them all together. This makes a larger number of features, but more detailed selections and overlays could be done with a subset of the buffer features. For instance, all the buffers representing the loudest noise could be totaled to show a percentage of the overall impact.

STUDY QUESTIONS

1. Could buffers be generated around a text field, such as a code?
2. How could you solve the problem of mismatched buffer value units and map units?

Other real-world examples

The public works department might buffer each street light in the city with a different value based on its wattage. This could provide a visual map of areas that are well lit and those that are not.

The local electrical utility might buffer its power lines by different values depending on the lines' voltage. This may be used to determine areas required for clearance easements and the value of the included property.

Tutorial 5-5

Using multiple buffer zones

The Straight-line Buffer command can be repeated several times to analyze several distances, but there is a tool that allows multiple distances to be entered for the buffering process.

Learning objectives

- *Create multiring buffers*
- *Perform distance analysis*

Preparation

- *Review page 126 in* The ESRI Guide to GIS Analysis, Volume 1

Introduction

The Multiple Ring Buffer tool allows the user to input several distances at once to create multiple buffers. On the surface, this looks like the regular Buffer tool runs multiple times, but there are some key features of the Multiring Buffer command that distinguish it.

When the multiring buffers are created, a dissolve option is presented. Not dissolving will make a full circle feature for each distance provided. The features will overlap. If these are used to do an overlay analysis, features could reside in several of the buffer rings created. In fact, features at the center of the buffered area would be inside all of the buffer rings.

If the output multiring buffers are dissolved, the features resemble a doughnut with the interiors clipped out where they overlap the other rings. Now, an overlay analysis would place the features in only one ring. They would be farther than one distance, but nearer than the next distance.

The other important difference is that the output buffer rings contain a field showing the buffer distance for each ring.

It's also important to note that the distance values can be of any sequence. They do not have to be at regular intervals.

Scenario　The fire chief wants you to find the number of houses within certain distances of each fire station. The chief figures that a fire truck averages about 30 mph on a fire run, and wants you to create rings that represent 1 minute, 2 minutes, and 3 minutes driving time from each fire station. You'll create those rings and do an overlay analysis on the building footprints.

Data   The chief provided you with the fire station locations. Each has a station number with this code:

      551 = Station 1

      552 = Station 2

      553 = Station 3

Building footprints were compiled from aerial photos taken last year, with a polygon representing each building. Usage codes, called UseCode, represent how each building is used:

      1 = Single Family Residential

      2 = Multi-Family Residential

      3 = Commercial

      4 = Industrial

      5 = Government

      6 = Utilities

      7 = Schools

      8 = Church

The parcel data is added for visual interest, but won't really have an effect on the analysis.

Create a multiring buffer

1   In ArcMap, open Tutorial 5-5.mxd. Shown are the parcel lines, the building footprints, and Station 1, symbolized by a blue box. You'll select the point representing the station and buffer it. By default, the Buffer tool will only buffer the selected features.

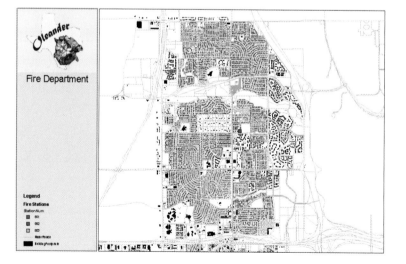

2   Using the Select Features tool , select the point representing Fire Station 1.

3   In the Search dialog, locate and open the Multiple Ring Buffer tool.

4   Set Input Features to Fire Stations.

5   Set Output Features to \GIST2\MyExercises\MyData.mdb and name the feature class **Station1Buffers**. The fire chief gave you an average vehicle speed and the times of 1, 2, and 3 minutes. Doing the math gives you distances of 2,640 feet; 5,280 feet; and 7,920 feet, respectively.

6   Enter the distance **2640** and click the plus sign to add the distance to the list.

7   Type in the distances **5280** and **7920**, clicking the plus sign after each to add them to the list. Verify that the buffer distance is set to feet.

8   Scroll down to the bottom and verify that the Dissolve Option is set to ALL. When your dialog box matches the graphic, click OK.

5-1
5-2
5-3
5-4
5-5
5-6
5-7
5-8
5-9

Select by location using buffer rings

The resulting buffer rings are added to the map document. The new layer needs to be at the bottom of the table of contents so that the other layers draw over the rings.

1 **Drag the new buffer layer below the Major Roads layer in the table of contents.** By using the Dissolve option, the rings should not overlap but be more of a doughnut shape. You'll check that by selecting the outer ring.

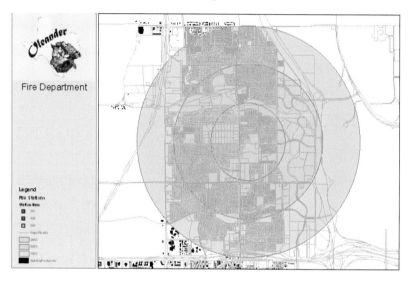

2 Clear the selected features and make the new buffer layer selectable. Using the Select Features tool, select the outer ring of the buffers. The significance of the doughnut ring is that you will be able to use it in your overlay processes to select features that are at least one distance way, but not farther than another distance. In this instance, you could select the housing units that are at least 5,280 feet away, but closer than 7,920 feet. You will also need to be careful to set your selection method so that houses that straddle the boundary don't get selected in two rings.

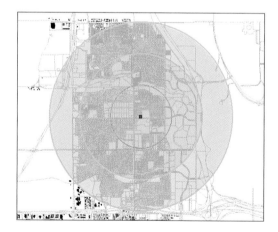

3 **On the main menu, click Selection > Clear Selected Features, or click the Clear Selected Features icon** ![icon] ▾ on the Tools toolbar.

4 With the Select Features tool, select the innermost ring of the buffers.

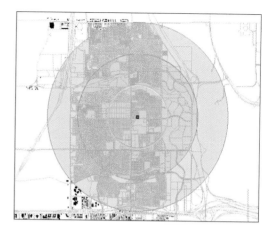

Once again, you are going to use one feature to select others with the Select By Location tool. You will want to select features from the building footprints that have some spatial relationship with the selected features of the Station 1 Buffers layer. But what selection method should be used? Take a look at the choices of relationships between the target layer and the source layer features:

Intersect	Features that straddle the boundary would be selected in two rings. (No)
Are within a distance of	We're using the buffers to determine near, not the selection. (No)
Contain	This is a reverse selection. Your structures would have to contain part of a ring. (No)
Completely contain	This is also a reverse selection. Your structures would have to contain an entire ring. (No)
Contain (Clementini)	Your structure would have to contain an entire ring without sharing the border. (No)
Are within	Any part of the building is inside one of the rings. (No)
Are completely within	Again, think of the features that straddle the boundary. (No)
Are within (Clementini)	Any part of the building is within the rings unless it shared the boundary. (No)
Are identical to	None of the structures will be the same as the rings. (No)
Touch the boundary of	Only the features that straddle the boundary would be selected. (No)
Share a line segment with	None of the structures will be coincident with the ring's arc. (No)
Are crossed by the outline of	Only the features touching the boundary will be selected. (No)
Have their centroid in	The centroid point can only occur in a single ring; no double selection. (Yes)

So after looking at all the choices, you will want to use the spatial selection method "Target layer features have their centroid in the Source layer feature." This will ensure that each building footprint is counted in only one ring.

5-1
5-2
5-3
5-4
5-5
5-6
5-7
5-8
5-9

5 On the main menu, click Selection >
Select By Location to start the process.
You've done this before, so try completing
the dialog box on your own. When it
matches the image, click OK.

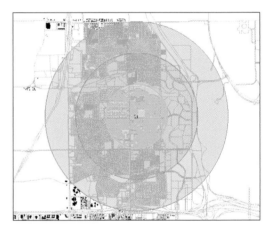

How many features were selected? How do
you know? Where can you find out?

6 Go to the List By Selection button at the top
of the table of contents. The number next
to "BuildingFootprints" is the number of
currently selected features. Make a note
of the number.

YOUR TURN

Perform the Select By Location process for the remaining two buffer rings. Make a note of the
total number of structures for each distance—you will use them in the next step.

7 Change the symbology of the Station1Buffers layer to Quantities > Graduated colors with distance as the Value field. Choose a light-blue-to-dark-blue color ramp, then flip the ramp so the zone nearest to the fire station has the darkest shade. Change the labels to reflect the number of structures in each buffer zone. Finally, move the buffer layer below the BuildingFootprints layer. When your dialog box matches the image, click OK.

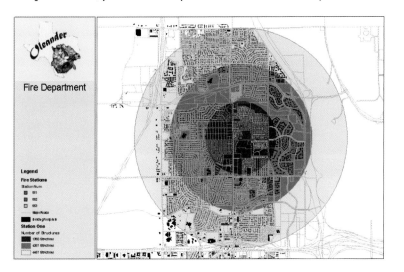

The blue shades were chosen to go along with the blue dot chosen for Station 1. Incorporating the selection results into the legend's color scheme keeps the map layout simple in form, yet still conveys the information clearly.

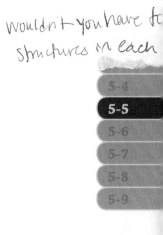

*Wouldn't you have t[o]
structures on each*

5-4
5-5
5-6
5-7
5-8
5-9

8 Save your map document as **Tutorial 5-5.mxd** in the \GIST2\MyExercises folder. If you are not continuing on to the exercise, exit ArcMap.

Exercise 5-5

The tutorial showed how to create multiple search distances with the Multiple Ring Buffer tool. Each ring became a selection feature for a Select By Location process.

In this exercise, you will repeat the process for the other two stations. For both Station 2 and Station 3, create a map that will display the number of structures in each of the three response times. For each set of buffers, use a color scheme that complements the corresponding station.

- Continue using the map document you created in this tutorial or open Tutorial 5-5.mxd from the \GIST2\Maps folder.
- Pan to the bookmark Station 2.
- Create the multiple ring buffers.
- Perform the selections for each ring.
- Set all the symbology for the layers, as well as any other title changes or notes, to make the map visually appealing.
- Repeat for Station 3.
- Save the results as **Exercise 5-5.mxd** in the \GIST2\MyExercises folder.

WHAT TO TURN IN

If you are working in a classroom setting with an instructor, you may be required to submit the maps you created in tutorial 5-5.

Turn in a printed map or screen capture of the following:

Tutorial 5-5.mxd

Two maps from Exercise 5-5.mxd

Tutorial 5-5 review

The multiple ring buffers are similar to the other buffers in that they are measured on a straight-line distance, and they create a feature whose area can be measured. They also give a dissolve option similar to the other buffers that can keep the buffer areas separate or cause them to overlap as necessary.

With any of the buffer tools, the final stages of the analysis become a simple inside–outside analysis that you learned about earlier. Once these buffer features are created, they can be used just like any other feature for overlays, or for more selections.

STUDY QUESTIONS

1. Explain the process of using multiple ring buffers to determine the exact areas of a soil layer that fell within each ring.
2. Could the same multiple ring buffer process be done with the regular Buffer tool or with the straight-line selection?

Other real-world examples

5-1
5-2
5-3
5-4
5-5
5-6
5-7
5-8
5-9

The local international airport might create multiple ring buffers around its runways. This could be used to set noise-mitigation zones that would determine the amount of sound insulation a nearby house must contain.

The fire department creates multiple ring buffers on the fly for large fires. These are used to set up restrictive zones to control who can enter or leave each zone.

Tutorial 5-6

Quantifying nearness

There are other common methods for finding what's nearby. These include the Near analysis, which calculates the straight-line distance from specific features to their nearest neighboring features, and spider diagrams, which visually display the distance from a specified feature to all the other features in the dataset.

Learning objectives

- *Calculate straight-line distance*
- *Draw spider diagrams*
- *Perform distance analysis*

Preparation

- *This tutorial requires two custom scripts to perform the spider diagram and Near analyses. They are both provided on the accompanying disk.*
- *Read pages 129–131 in* The ESRI Guide to GIS Analysis, Volume 1.

Introduction

Many of the tools introduced so far really only indicate if a feature is closer than or farther than a given distance from a feature. They do not answer the question of exactly how far it is from the feature. There are two processes that you'll deal with that will help to quantify nearness.

The first process is the Near analysis. A point is given as the central feature, and a distance to all the other features is calculated. The distances are stored in the attribute table, and analysis can be done with an exact distance quantity for each feature. The features may be color-coded by the distance from the source or used to identify the closest source.

The second process uses the spider diagram generator. Similar to a Near analysis, a central point is given, but this time a line is generated from that point to every other feature. Since lines automatically get a Shape_Length field, the distance is calculated, too. Straight-line travel distance can easily be displayed for visual analysis.

Scenario   The fire chief has provided some response data from last month and would like you to determine how many calls were answered by a crew from a station that was not the closest to the scene. From this information, the chief may be able to spot some understaffing or problems with the dispatch system.

Data   The chief provided the fire response data. It includes a large number of fields, including the addresses used to geocode the locations. You'll use the Station field to determine which station responded to each call.

This tutorial also uses two scripts for the Near process and the spider-diagram process. The ArcGIS Near tool needs an ArcInfo license and provides much more capability than the script provided here. The Point Near tool (derived from a script written by the author for this tutorial) is sufficient to demonstrate the Near analysis process.

The tutorial also uses the Create Spider Diagrams tool from the ArcScripts catalog (arcscripts.esri.com), written by Anthony Palmer of the U.S. Army Corps of Engineers. It allows many varieties of spider diagrams to be made from the single script.

Use the Point Near tool

1 In ArcMap, open Tutorial 5-6.mxd. The map displays the locations of all the fire calls that month. Notice that some went outside of the city limits. Oleander has a mutual-aid response agreement with all the neighboring cities to provide emergency service regardless of city limits.

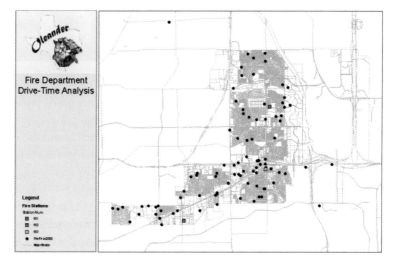

5-1
5-2
5-3
5-4
5-5
5-6
5-7
5-8
5-9

Included on the accompanying DVD is a script that will perform a distance analysis. It takes the input points and finds the nearest features from another specified layer. The script is in a custom toolbox, which you can navigate to and use.

2 Open the Catalog window and navigate to C:\ESRIPress \GIST2\Toolboxes. Click the GISTutorialTools.tbx toolbox to see the scripts it contains.

The method for analysis will be to run the Point Near tool and determine which station is closest for each call location. Then you'll match that number to the actual responding station to determine which ones had responders traveling too far.

3 Double-click the Point Near script to run it. Use the drop-down box to set Input Features to FireRuns0505. Set Near Features to Fire Stations. When your dialog box matches the graphic, click OK to run the tool.

4 A pop-up box will appear when the script has completed. Click it to see the messages. Close the box when you have examined the messages. If you miss the pop-up box, the results window can be opened by clicking Geoprocessing > Results.

5 Right-click FireRuns0505 layer and select Open Attribute Table. Move the slider bar all the way to the right to see the new fields that have been added to the table. The NEAR_DIST and NEAR_FID attributes have been added to the table identifying which station is closest and the straight-line distance to that station. The Feature ID from the Fire Stations table was transferred to the FireRuns0505 layer, and you'll have to join those tables to determine the station number for comparison.

Join tables

1 Close the attribute table. Right-click the FireRuns0505 layer and select Joins and Relates > Join.

2 Set the Join type to "Join attributes from a table."

3 Set the join field to Near_FID, which was established by the Point Near tool.

4 Set the Join table to Fire Stations.

5 Set the field in the join table to OBJECTID. When your dialog box matches the graphic, click Validate Join. A list of validation checks for the join will be displayed.

6 If the join passes the validation, close Join Validation and click OK to perform the join. Otherwise, recheck your settings and validate the join again.

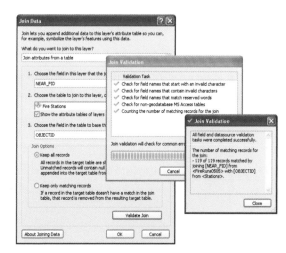

7 Open the attribute table for the FireRuns0505 layer again and scroll to the far right.

The Join procedure has added the attributes from the Fire Stations layer to the Fire Runs layer. Now you can do some checking to see the cases in which the responding station was not the nearest station. You'll do that with a definition query, looking for features where the station numbers are not equal.

8 Close the attribute table. If you remember from the previous tutorials, a definition query is used to limit the data that's shown in a feature class or table. While it doesn't delete any data, it makes subsets of the data totally invisible to the user.

9 Right-click the FireRuns0505 layer and open Properties. Go to the Definition Query tab and click the Query Builder button. Enter the query **FireRuns0505. STATION <>Stations.StationNum**. When your dialog box matches the graphic, click OK and then OK again to apply the query.

5-1
5-2
5-3
5-4
5-5
5-6
5-7
5-8
5-9

YOUR TURN

Set a unique values classification on the FireRuns0505 layer using the field
STATION. Set the colors to match the existing colors for the Fire Stations layer.
Adjust the symbol size as necessary.

Station 551   Cretean Blue
Station 552   Fire Red
Station 553   Quetzel Green

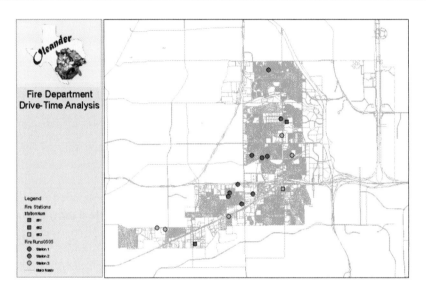

This map shows the calls that did not originate from the nearest station and is ready for
the fire chief to use for some quick visual analysis. The incidents may be investigated
further to determine any nongeographic reasons for the problems.

One thing the map doesn't show is the other calls that did originate from the nearest
station. Perhaps adding a spider diagram to the map will clarify the correct call responses.

10 Right-click the FireRuns0505 layer. Select Joins and Relates > Remove Joins > Stations to
remove the join. Note that the definition query will automatically be removed since it
depended on fields from the join.

Create a spider diagram

A spider diagram will draw a line from each response location to the fire station where the call originated. The mass of lines gives the viewer an overall pattern of distance and direction. The tool to create a spider diagram is available on the ArcScripts Web site or on the disk included with this book.

1 Open the Catalog window again and locate the GISTutorialTools.tbx toolbox. Double-click Create Spider Diagrams.

2 Click the drop-down menu for Origin Feature Class and select FireRuns0505. Set the Origin Key Field to STATION.

3 Set the Destination Feature Class to Fire Stations. Set the Destination Key Field to StationNum.

4 Finally, set the Output Feature Class to \GIST2\MyExercises\MyData.mdb and name the feature class **FireSpider**. When your dialog box matches the graphic, click OK to start the process.

The Spider Diagram tool will draw a line from every feature in the FireCalls0505 layer to each fire station. In order to get the correct view of the data, you need to look only at the lines where the ORIG_ID and the DEST_ID are the same. The lines where these two values are not equal do not represent real calls, but were made in the blanket operation of the Spider Diagram tool.

To restrict the data, you'll set up a Definition Query where ORG_ID = DES_ID. This will match the calls to their originating station.

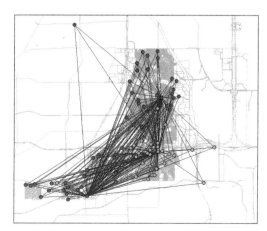

5 Right-click the FireSpider layer and click Properties. Go to the Definition Query tab and build the query **[ORG_ID] = [DES_ID]**. Click OK, and OK again to accept the query.

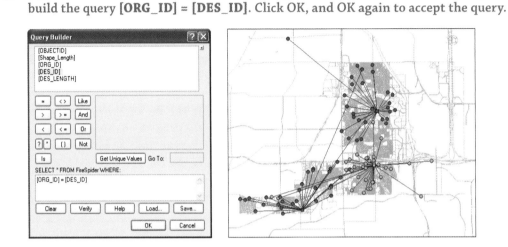

YOUR TURN

Set the symbology for the new FireSpider layer to unique values using DES_ID as the Value Field. Color each value to match the color of the station symbols they are linked to. Adjust the line thickness as necessary.

The final map shows all the calls along with a spider diagram color-coded to match the responding station. We're giving the fire chief a little more than asked for, but this will give more information for visual analysis.

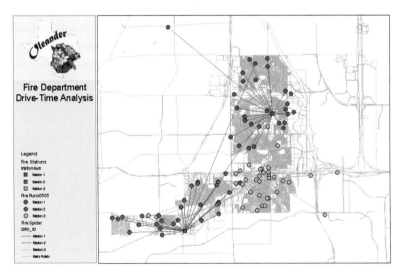

6 Save your map document as **Tutorial 5-6.mxd** in the \GIST2\MyExercises folder. If you are not continuing on to the exercise, exit ArcMap.

Exercise 5-6

The tutorial showed how to find the nearest feature to a set of input points. The Point Near tool calculated the nearest feature, as well as storing the distance to that feature. In addition, the Spider Diagram tool creates a linear feature from each point to its nearest point.

In this exercise, you will repeat the process using a different dataset. The tutorial used the fire department calls for service, showing the relationship between the call location and the fire station from which the call was dispatched. For this exercise, you will create a similar map using ambulance calls and analyze the relationship between the call location and the fire stations from which the ambulances were dispatched.

- Continue using the map document you created in this tutorial or open Tutorial 5-6.mxd from the \GIST2\Maps folder.
- Turn off the FireRuns0505 and FireSpider layers.
- Add the AmbulanceRuns0505 layer from the \GIST2\Data\CityOfOleander\Fire Department feature dataset.
- Run the Spider Diagram tool to create the same analysis as you did with the fire run data.
- Set all the symbology for the layers, as well as any other title changes or notes to make the map visually appealing.
- Save the results as **Exercise 5-6.mxd** in the \GIST2\MyExercises folder.

5-1
5-2
5-3
5-4
5-5
5-6
5-7
5-8
5-9

WHAT TO TURN IN

If you are working in a classroom setting with an instructor, you may be required to submit the maps you created in tutorial 5-6.

Turn in a printed map or screen capture of the following:

> **Tutorial 5-6.mxd**
> **Exercise 5-6.mxd**

Tutorial 5-6 review

The Near function can be used to quantify a discrete feature's proximity to another location. The tool is available in ArcInfo, but ArcView and ArcEditor users can use the provided custom tool to mimic the function. With buffers, all you knew was that a certain location was farther away than one distance but closer than another. With Near, an exact distance is recorded. It's important to note, however, that a point must exist at the location for which you want the distance value.

The Spider Diagram tool is almost always used for flashy visuals. The distance for each spider line is calculated, but because they are all straight-line distances their usefulness is limited. Leaving all the spider lines visible gives a good graphic of the total service area for the data. The fire department data you worked with gave you the ability to separate each station and link them only to calls they made, giving you three service areas, one for each station.

STUDY QUESTIONS

1. When would you show spider lines for all origins and all destinations, and when would you separate them?
2. What is the major drawback of the Point Near tool?

Other real-world examples

A spider diagram might be created between known bird nests and sightings of an endangered species. This would represent the travel territory of the birds.

The Near command could be used with business locations and public transit stops to determine which is closest, next closest, and so on.

Insurance companies may use the Near value of a house to fire hydrants to help set rates.

Tutorial 5-7

Creating distance surfaces

Straight-line distance analysis can also be done with raster data. The values are formed into a continuous coverage, with each pixel containing the distance from the subject feature. The process can include a cost to traverse the distances, resulting in a cost-distance surface.

Learning objectives

- *Work with raster data*
- *Calculate costs*
- *Perform distance analysis*

Preparation

- *This tutorial requires the ArcGIS Spatial Analyst extension.*
- *Read pages 132–134 and pages 142–147 in* The ESRI Guide to GIS Analysis, Volume 1.

Introduction

The ArcGIS Spatial Analyst extension allows for many of the same types of analysis that you do with vectors to be done on raster images. The advantage is the ability to overlay a large number of layers in one process and to have the process complete quickly.

The distances you calculated for the discrete point data were only for those points. Any area that did not have a point did not get a distance value calculated. With a raster image file, the data will be continuous across the entire study. So even if there are not discrete points at a location, a distance value will be stored creating a continuous distance surface. In raster analysis, it would be possible to overlay this distance surface with data such as slope to find areas of a certain slope and distance value combined.

The distance surface can also include a cost to traverse the surface. This will add another factor to the analysis, letting you investigate other influences to the nearness of features.

Scenario   Using the data from the fire chief, you need to create a distance surface that can be used as a backdrop for some of the data analysis. You can do a very smooth gradation of color over the entire distance, which might add some impact to your map.

Data   The Fire Stations layer will be used to set the starting locations for the distance surface. The fire-calls data will let you do some quick visual analysis for distance.

Create a raster ring buffer

1 In ArcMap, open Tutorial 5-7.mxd. Here are the fire calls and station locations again. You'll do the distance analysis for Station 1 by making a single distance surface that goes out three miles. Next, you'll do Station 2 and see what areas are within one mile of both stations.

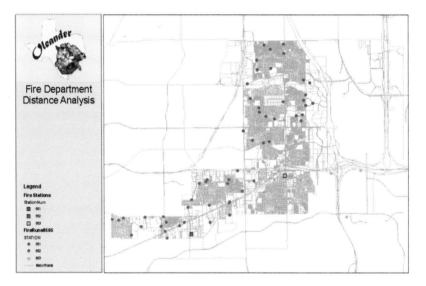

2 On the main menu, select Customize > Extensions. If necessary, check the box next to Spatial Analyst. If this does not appear in your list, or an error message responds that Spatial Analyst is not available, see your system administrator.

3 Check to make sure that Fire Station 1 is selected. If it is not, use the Feature Selection tool to select it.

You will also need to set a processing extent. By default, Spatial Analyst will use the extent of the input file which will almost always work for raster files. But your input file will be the point locations of the fire stations, meaning that the default extent would be the smallest box necessary to enclose all of the points. The result would not cover the entire city, so you will need to set an extent in the geoprocessing environment that does cover the entire city.

4 On the main menu, select Geoprocessing > Environments. Click the Processing Extent arrow to open the input dialog box. Change the Extent to Same as layer Lot Boundaries. Click OK.

5 Open the Search window and locate the Euclidean Distance tool. Click to open it.

6 Use the drop-down arrow to set the Input raster or feature source data to Fire Stations. Set the Output distance raster to the MyData geodatabase and name it **Station1_Distance**. Set Maximum distance to **10560** (2 miles) and Output cell size to **50**. When your dialog box matches the graphic, click OK.

The result is a bull's-eye layer highlighting the distance from the fire station. The default classification uses just a few colors, making the rings look a little rough. You'll smooth that out later when the final version is done.

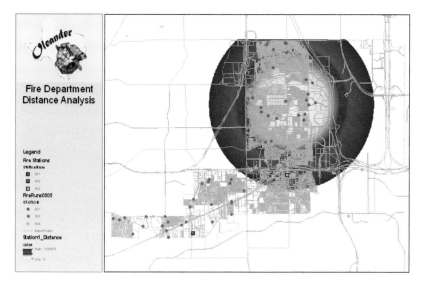

5-1
5-2
5-3
5-4
5-5
5-6
5-7
5-8
5-9

YOUR TURN

Clear the selected features and select Fire Station 3 (the green symbol), then repeat the process. Make a distance surface using the distance of **10560** and a cell size of **50**. Save it as **Station3_Distance**.

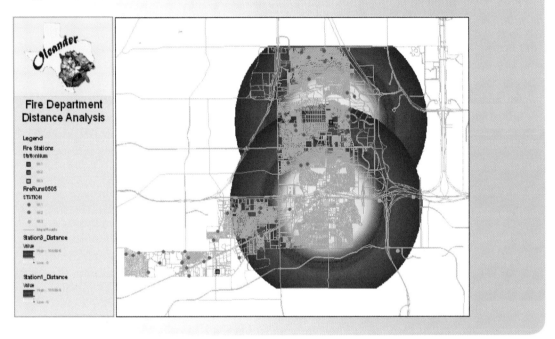

This will make a double bull's-eye. You're going to add these two rasters together, which means that at each pixel the value for each layer will be added together. Locations with a close proximity to both stations will get a high value, and those farther away will get a lower value. You'll use the Spatial Analyst tool Cell Statistics to add the cell values together.

Combine rasters

1 In the Search window, click the Cell Statistics tool to open it.

2 Set the inputs as both Station1_Distance and Station3_Distance. Make the output file **Station1_Plus_3_Distance** in the MyData.mdb geodatabase. Set the Overlay statistic to SUM. When your screen matches the graphic, click OK.

3 Turn off the two bull's-eye layers you created before, leaving the new layer visible. The default classification is a gray scale with the lowest values being the darkest. Change this to a red palette with the highest values being the darkest.

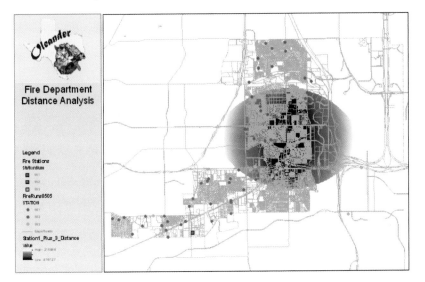

4 Right-click the Station1_Plus_3_Distance layer and open the Properties dialog box. On the Symbology tab, click the Color Ramp drop-down menu and select a monochrome red scale with the darkest value on the right. Click OK. The red zone shows relative distance from each fire station. The darker areas are farther away, making them more of a concern for the fire chief.

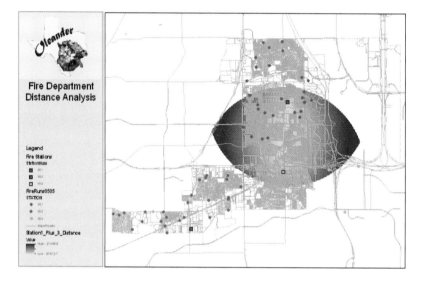

5 Save your map document as **Tutorial 5-7.mxd** in the \GIST2\MyExercises folder. If you are not continuing on to the exercise, exit ArcMap.

Exercise 5-7

The tutorial showed how to create distance surfaces, then combine the results into a single raster by doing a cell-by-cell summation. The result was a continuous phenomenon dataset that displayed distance from the fire stations.

Before you take this map to the fire chief, you need to perform the surface distance analysis process for the intersections of Station 2 and Station 3.

- Continue using the map document created in this tutorial or open Tutorial 5-7.mxd from the \GIST2\Maps folder.
- Create a distance surface for Station 2 (and for Station 3 if you didn't do it in the tutorial).
- Add the surface for Station 2 to the surface for Station 3 with the Cell Statistics tool.
- Adjust the legend and symbology for the new surface.
- Save the results as **Exercise 5-7.mxd** in the \GIST2\MyExercises folder.

WHAT TO TURN IN

If you are working in a classroom setting with an instructor, you may be required to submit the maps you created in tutorial 5-7.

Turn in a printed map or screen capture of the following:

> **Tutorial 5-7.mxd**
>
> **Exercise 5-7.mxd**

5-1
5-2
5-3
5-4
5-5
5-6
5-7
5-8
5-9

Tutorial 5-7 review

The distance surface commands are practically the same thing as the Near command, except that a value is calculated for every location on the map. Using the Identify tool, you could click anywhere on the map and get a distance value.

The distance surface also required you to do two steps to get a similar result as the Near command. You could only create a distance surface for one station at a time. Then you had to add the two rasters together to get a value that represented the distance to the nearest fire station.

The distance surface also suffers from the same major drawback as the Near command: the distances are all straight line (Euclidean). Things like lakes and freeways are not considered to be obstacles in getting from one place to another.

STUDY QUESTIONS

1. How are the raster values different from the point values that were created with the Point Near command?

2. What data is stored in a density surface?

3. When would the Euclidean distances cause problems?

Other real-world examples

A radio station may create a distance surface to determine a reception area. The overlapping areas might be used to investigate the impact of the radio waves on animal habitat.

When doing raster analysis, the distance surface can be used to introduce a distance element to the formula. A distance surface may be generated around a body of water and added to the raster data for crop health to investigate possible contamination sources.

Tutorial 5-8

Calculating cost along a network

Analysis using straight-line distance has an inherent flaw when dealing with streets or other types of networks. The distance measurements may cross areas that the network doesn't go. The solution is to use Network Analyst tools to calculate a cost along the network.

Learning objectives

- *Build a network*
- *Calculate costs*
- *Perform distance analysis*

Preparation

- *This tutorial requires the ArcGIS Network Analyst extension.*
- *Read pages 135–141 in* The ESRI Guide to GIS Analysis, Volume 1.

Introduction

All of the tutorials you've done with fire stations so far involved measuring a straight-line distance (Euclidean distance, or "as the crow flies") from the analysis location points. The problem with this is that sometimes there are places where the streets just don't go, like across lakes or other natural barriers.

The Network Analyst tools let you overcome some of these difficulties by doing distance analysis along the network. Networks can be streets, pipelines, electric transmission lines, creeks, train tracks, and so on. These networks all have connectivity and something traverses them. To get the most accurate analysis, you assign a cost for traversing each segment of the network. If you have cost per mile for each street segment, the Network Analyst extension could show how far you could get for $1.00. Electric transmission lines might be limited by capacity in volts, or a pipeline might be restricted by flow rate based on pipe size.

For this tutorial, you'll calculate the time it takes to traverse each segment and see how far you can get in a specified number of minutes. This is more accurate than a distance-only cost because it takes into account speed limits. Note that although you're using time as the cost, the process will work equally well with distance only, if speed limit data is not available.

Scenario   It seems that anytime you do a great map, the fire chief suddenly thinks of more maps to request. After analyzing the previous maps, the fire chief has seen that Station 1 is being covered by Station 3 quite a lot. It seems that the new development at the north end of town is calling out the Station 1 crew more and more, leaving the rest of its district to be covered by others. What you couldn't see in your analysis is the number of times adjacent cities backed up Station 1 and came into Oleander on fire calls under the mutual-aid agreement.

The chief wants to start the process with the city council to approve a new station. Part of the justification is to show that there are areas not being covered in the standard response time. You'll use the Network Analyst extension to show how far the fire trucks can get in a given amount of time.

Data The street centerline file will be used for the network analysis. The fields SpeedLimit and Shape_Length will be used to calculate the time it takes to traverse each line segment.

The FireStations layer will be used to set the starting locations for the analysis.

Create a network database

1 Start ArcMap, open Tutorial 5-8.mxd, and activate the Network Analyst extension. Click the Catalog tab on the right of the map document, navigate to the directory where you installed the tutorial data, and expand the Data folder. Right-click the Networks.mdb geodatabase and select Copy.

2 Scroll down in the Catalog tree and find the MyExercises folder. Right-click it and select Paste. This will give you a copy of the data to use for the network, in case things get messy. It will also provide you with a place to store all of the analysis results.

You'll need to add a field to the Street Centerlines feature class to store the time data you are going to calculate.

3 Expand the MyExercises folder, the Networks geodatabase, and the Network Analysis feature dataset. Right-click StreetCenterlines and select Properties.

4 In the Properties dialog box, go to the Fields tab. Scroll all the way to the bottom of the fields list and click in the first open line. Type in a new field name—**MINUTES**—and set the Data Type to Double, then click OK.

You'll create a network using this new field as a cost, which will let you do routing based on the number of minutes it takes to complete the route. The field is empty now, but you'll populate it with values later.

5 Right-click the feature dataset Network Analysis and select New > Network Dataset to start the creation process.

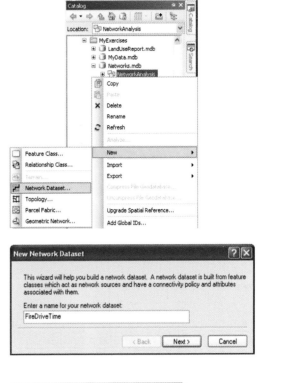

6 In the first screen of the New Network Dataset wizard, name the new dataset **FireDriveTime**.

7 Click Next. Click the box to add StreetCenterlines to the network dataset.

8 Click Next. Accept the defaults for the next six screens, clicking Finish to create the network dataset.

The important screen in the process was setting your new field to be the cost, but since you used the default keyword of Minutes, it will automatically be used for the cost field. Other keywords that will be recognized as a cost for a network can be found in the file NetworkDatabaseConfiguration.xml, which loads with the Network Analyst extension.

9 When prompted to build the network and add it to the map document, answer No. You'll add it to the document and build it later when the MINUTES field is populated.

Set up the network

1 Close the Catalog window, if necessary, to reveal the map. The map looks similar to the one in the previous tutorial, displaying the fire department calls for service and the street centerlines. You need to add the new street network you just built, calculate the MINUTES field for each line segment, and identify the fire stations as the starting points for the analysis.

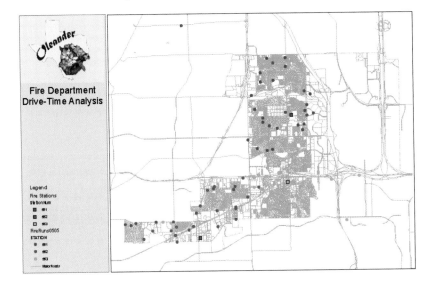

2 Click the Catalog tab to open the Catalog window. Find the FireDriveTime network and drag it to the top of the list in the table of contents.

3 When prompted to add all the feature classes that participate in the FireDriveTime network, click Yes.

4 Right-click the StreetCenterlines layer and select Open Attribute Table. Scroll the table all the way to the right and find the MINUTES field.

5 Right-click the MINUTES field and select Field Calculator. If necessary, click Yes on the warning box. Build the equation [Shape_Length] / ((([SPEEDLIMIT] × 5280) / 60).

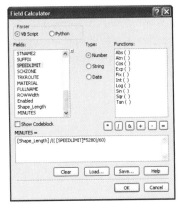

6 Click OK. The result is the time in minutes that it takes to traverse each line segment. Close the attribute table.

Perform network analysis

1 Turn on the Network Analyst extension in ArcMap, if necessary. Add the Network Analyst toolbar to ArcMap by clicking Customize > Toolbars > Network Analyst on the main menu. Dock it if you wish. The toolbar will by default display the FireDriveTime network since it is the only one that exists in the table of contents. Now that you've populated the MINUTES field with the time it takes to traverse each line segment, you can build the network.

5-1
5-2
5-3
5-4
5-5
5-6
5-7
5-8
5-9

2 Click the Build Network Dataset button 🕸 . This network will now let you do several types of network analysis, including routing, finding the nearest facility, an origin–destination cost matrix, and the process we're going to use: service area.

The Closest Facility function is similar to the Point Near tool you used earlier, except that it measures distance along the road instead of the straight-line distance. You'll do that in the next tutorial. The Origin–Destination (OD) Cost Matrix function produces similar results as the Spider Diagram tool that was used earlier.

The first step is to add the framework in which your analysis will take place. To do this you'll use the New Service Area tool.

3 On the Network Analyst toolbar, click Network Analyst and select New Service Area. This adds a new Group Layer to the table of contents called Service Area, along with a series of feature classes that will be used to hold the components of your analysis. These are currently empty, but they will be populated as you work with the network.

The Facilities layer will hold the starting locations for your service area analysis. These will be the fire stations. You could place them into the feature class manually using the Create Network Location tool on the Network Analyst toolbar, which wouldn't be too bad for the three stations. But if you had more stations, then an automated tool would need to be used. You'll go ahead and use the automated tool to add the locations.

4 Use the Search window to locate and open the Add Locations tool.

Even though you have the fire stations as a feature class in your map, they need to be matched up with the network database that you made. The Add Locations tool will move them into the framework of the network so that they can be used in the analysis.

5 In the drop-down menu, set the Input network analysis layer to Service Area and set the Sub layer to Facilities.

6 Set Input locations to Fire Stations. In the Field mappings pane, set the Name field to **StationNum**.

7 Scroll to the bottom of the dialog box and check the Snap to Network box. When your dialog box matches the graphic, click OK.

If you were to zoom to the fire stations, you would notice a new dot placed there by the Add Locations tool. This point is now part of the network database and can be used with the various network tools. If you noticed, one of the other Sub Layers was Barriers. These are elements that you place on the street network to prevent the analysis from going through that point. For instance, if a road were closed for repair, you could place a barrier at either end, and that road segment would not be used in routing calculations.

Now that you have the fire stations added as network facilities, you can set the parameters for the service area calculation. You'll need to define the measurement units you are going to use and the distance from the fire stations you want your analysis to extend.

The fire chief wants to see service areas for total response times of three, four, and five minutes. The standard emergency response allows sixty seconds to dispatch the call to a station, and another sixty seconds to get the firefighters on the trucks and out the door. Subtracting that two minutes from the total response times gives drive times of one, two, and three minutes to reach the location.

8 Right-click Service Area in the table of contents and select Properties.

9 In the Properties dialog box, go to the Analysis Settings tab. The Impedance is already set to Minutes since ArcMap recognized the keyword as a valid field name. Set Default Breaks to **1**, representing 1 minute. When your dialog box matches the graphic, click OK. Everything is set up for the analysis; the facilities are loaded and the service area parameters are set to 1 minute.

10 On the Network Analyst toolbar, click the Solve icon ▦ . Polygons have been created showing the service areas for each station. Notice also that they are not the nice, rounded buffers you got before. This is a good example of the difference between straight-line distance and distance along a network.

11 Turn off the street centerlines and the FireDriveTime_Junctions layers to make your map easier to read.

5-1
5-2
5-3
5-4
5-5
5-6
5-7
5-8
5-9

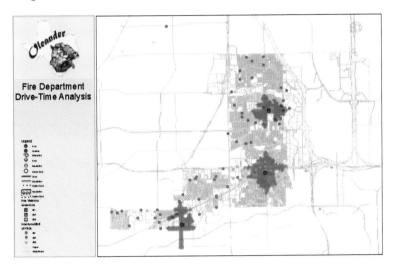

The polygons created in the drive-time analysis are temporary and will go away if you change the parameters and solve for service areas again, so you need to export the results to a new feature class.

1 In the table of contents, right-click the Polygons layer and select Data > Export Data.

2 Click the Browse button and set the Output Feature Class to **1_Minute_Service_Areas** in \MyExercises\MyData.mdb. Click Save and OK. When prompted, elect to add the new layer to the map document.

YOUR TURN

Change the symbology of the 1_Minute_Service_Areas layer to make it more presentable:

- Set a unique value classification using the Facility ID.
- Color-shade each facility according to the colors used before for the stations.
- Set the transparency to 50 percent.
- Move the 1_Minute_Service_Areas layer above Street Centerlines in the table of contents.

You may want to uncheck the Service Area group layer to check your work, but turn it back on before proceeding.

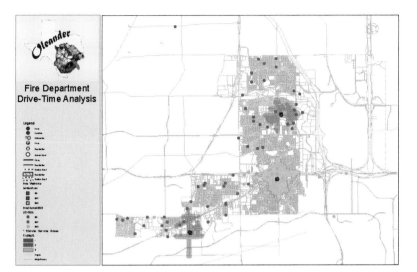

Now that you have mapped the one-minute service areas, you need to do the two- and three-minute service areas. To do this, you'll change the parameters of the Service Area layer and Solve again.

3 Turn off the one-minute service area layer. Right-click the Service Area layer and select Properties. Go to the Analysis Settings tab and change Default Breaks to **2**. Click OK.

4 On the Network Analyst toolbar, click the Solve button. The new service areas are generated, and you see much more coverage.

5-1
5-2
5-3
5-4
5-5
5-6
5-7
5-8
5-9

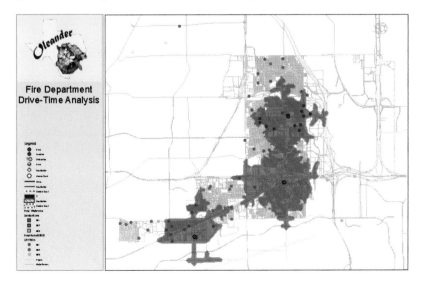

YOUR TURN

Export the Polygons 2 layer to a new feature class called 2_Minute_Service_Areas, and add it to the map document.

Change the symbology of the new layer to make it more presentable:

- Set a unique value classification using the Facility ID and arrange the colors to match the one-minute drive time (**Hint:** On the Symbology tab, click the Import button and use the settings from the one-minute layer).
- Set the transparency to 60%, lighter than the other layer
- Repeat the whole process to create the three-minute service areas.
- Set the colors as before and set the transparency to 80%.
- Review the order of layers and move them in the table of contents to make the map clear and concise.
- Turn off the Service Area layer, which will turn off all its associated layers.

The map now shows the one-, two-, and three-minute response zones, along with a few areas that don't seem to have very good coverage. Perhaps this will help the case for funding a new fire station.

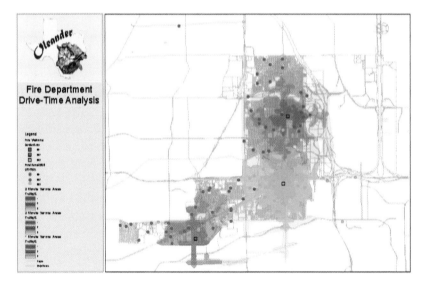

In this tutorial you created each of the response zones separately. However, you could have done it in one step. The Analysis Settings dialog will accept multiple costs separated by a space, and other settings let you decide between overlapping polygons and ring polygons. Explore these options to see how they may be used for future projects.

5 Save your map document as **Tutorial 5-8.mxd** in the **\GIST2\MyExercises** folder. If you are not continuing on to the exercise, exit ArcMap.

Exercise 5-8

The tutorial showed how to establish a service area based on the cost of traveling through a network. The "cost" was time and the "network" was the streets.

The fire chief was impressed by the map and asked if you could count the number of buildings in each of the three service areas for each time value. The results of the service area analysis are polygons, and with building footprints available, it becomes a simple "finding what's inside" task. The chief will want the map when it's finished.

- Continue using the map document you created in this tutorial or open Tutorial 5-8.mxd from the \GIST2\MyExercises folder.
- Add the layer BuildingFootprints from the City of Oleander geodatabase, Planimetric feature dataset.
- Use each of the three service areas to determine the number of buildings in each.
- Add a textbox to the map title area to show the results.
- Set all the symbology for the layers, as well as any other title changes or notes to make the map visually appealing.
- Save the results as **Exercise 5-8.mxd** in the \GIST2\MyExercises folder.

5-1
5-2
5-3
5-4
5-5
5-6
5-7
5-8
5-9

WHAT TO TURN IN

If you are working in a classroom setting with an instructor, you may be required to submit the maps you created in tutorial 5-8.

Turn in a printed map or screen capture of the following:

Tutorial 5-8.mxd

Exercise 5-8.mxd

Tutorial 5-8 review

Finally the problem of straight-line distance analysis is solved. Measurement along the streets is a better representation of travel time than the "as the crow flies" model of the buffer and raster surface tools.

There was a lot of data setup in order to get the Network Analysis tools to work. A good street centerline file with perfect connectivity was needed, along with a way to calculate some cost along the line. In many cases this is merely distance, but this would not give priority to freeways where you can drive much faster. Your dataset had speed limits, which let you calculate the traverse time for each line segment. It is also important to be aware of one-way streets and restricted turns. Your database didn't account for these rules, but a more complex dataset could have these characteristics built in.

Remember that a network is only a model of reality. You could never really build in the true complexity of a real street network. Why don't the Internet mapping programs produce the exact route you take to work every day? Very simply, your observational model is more complex. You know when traffic will be heavy, in what condition certain streets are, and the reality of making some of the more difficult turns. The computer models are much more generalized, although the ability to make them very complex does exist.

STUDY QUESTIONS

1. At what point will the accuracy of your network model negatively affect your results?
2. How could a model based on speed limits be more accurate?
3. You only did drive-time analysis on this model, but it is capable of routing. Using the Network Analyst toolbar, try adding some stops and an origin. Are the routes created by the Network Analyst tools acceptable?

Other real-world examples

A flower delivery service might create a drive-time polygon around each of its member stores to show what areas are serviceable and which store would deliver to a particular area.

The newspaper delivery office may use routing to determine the best path for its delivery agents, given a starting point and a list of delivery locations.

The police department's criminal investigation unit might create a drive-time polygon to see if a suspect could drive from his location to the crime scene in fifteen minutes.

Tutorial 5-9

Calculating nearness along a network

The Near command from the general ArcToolbox analysis toolset uses a straight-line calculation, which is not always the most accurate method. Using Network Analyst, the Near command can be duplicated, but with the results being measured along the network.

Learning objectives

- *Calculate proximity along a network*
- *Perform distance analysis*

Preparation

- *This tutorial requires the ArcGIS Network Analyst extension.*
- *Review pages 129–131 in* The ESRI Guide to GIS Analysis, Volume 1.

Introduction

You solved the "as the crow flies" dilemma by building a network and using it for analysis. Another tool in Network Analyst mimics the Near tool you used previously. The Near tool measured the straight-line distance from a source location to every given point. The Closest Facility tool will run the same analysis, but give the results as the cost along a network.

As stated before, this mitigates anomalies with the data such as having to drive around lakes or closed roads.

Scenario   The fire chief wants the Near analysis that you did before updated to follow the network. Several of the city council members remarked that the map had some calls going right through the high school football stadium, and that can't be right. And while it is "right" for the tool you used, there is another way to calculate a near distance that will be closer to reality.

You will repeat the analysis, but this time use the New Closest Facility tool in Network Analyst.

Data   The Network database built in the previous tutorial will be loaded again, with minutes being the cost of traversing each road segment.

The Fire Stations and FireRuns0505 layers will be used for the origins and destinations in the analysis.

Determine nearness using driving distance

1 In ArcMap, open Tutorial 5-9.mxd. This should look familiar, as it is the fire response calls and fire station locations. You'll need to add to the map document the network database that you built; then you'll be able to start the analysis.

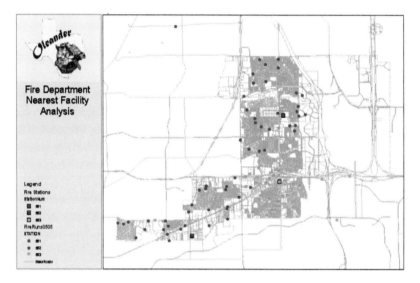

2 Click the Catalog tab and navigate to the FireDriveTime network you created in tutorial 5-8. Drag it to the top of the layer list in the table of contents. When prompted, add the other feature classes that participate in the network. Turn off the FireDriveTime_Junctions layer.

3 If your Network Analyst toolbar is not visible, add it now. On the Network Analyst toolbar, click Network Analyst and select New Closest Facility. This will add a set of layers to contain the components of the analysis.

You will need to add Facilities and Incidents to the layers as analysis locations. The Facilities will be the fire stations, and the Incidents will be the fire calls.

4 Use the Search window to locate the Add Locations tool and open it.

5 Set the Input network analysis layer to Closest Facility and set the Sub layer to Facilities.

6 Set Input locations to Fire Stations and the Name field to **StationNum**.

7 Scroll to the bottom of the dialog box and check the Snap to Network box.

8 Click OK to run the tool.

YOUR TURN

Repeat the process to add the fire calls as Incidents.

- Run the Add Locations tool.
- Set the Input analysis layer.
- Set the Sub layer to Incidents.
- Set the Input locations to FireRuns0505.
- Set the tool to Snap to Network.

Everything should be set for the analysis. When you start the process, a route along the street network will be calculated from each fire station to every response call.

9 On the Network Analyst toolbar, click the Solve button 🏁 .

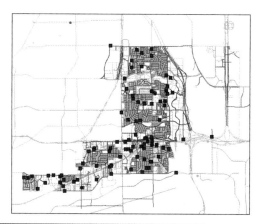

YOUR TURN

Just as in the last tutorial, the results of the analysis are temporary. You'll need to save them to another feature class. Then you'll want to symbolize the routes to represent the nearest station.

- Export the Routes layer as **NearNetwork** in the MyData.mdb geodatabase.
- Symbolize the new layer using the Facility ID field.
- Turn off the Closest Facility layer group, Street centerlines, and FireDriveTime.
- Review the order of layers and labels to make the map clear and concise.

This completed map should be just what the fire chief wanted. Now each route is shown in the color of the responding fire station. The situations where the incident color does not match the route color indicates that the responding station was not the closest.

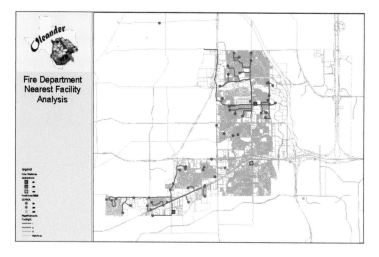

10 Save your map document as **Tutorial 5-9.mxd** in the \GIST2\MyExercises folder. If you are not continuing on to the exercise, exit ArcMap.

Exercise 5-9

The tutorial showed how to use the network routing to find the nearest facility to a set of incidents using time.

The fire chief would also like to see the results of the near along a network analysis for the ambulance calls. The process will be the same as that used for the fire response calls but using this different dataset. Repeat the process using the ambulance response data you used previously.

- Continue using the map document you created in this tutorial or open Tutorial 5-9.mxd from the \GIST2\Maps folder.
- Add the layer AmbulanceRuns0505 from \GIST2\Data\City Of Oleander.mdb\ FireDepartment.
- Add the Network Database for the streets that you created in this tutorial, if necessary.
- Run the Closest Facility analysis.
- Set the Fire Stations as Facilities and the Ambulance Calls as Incidents.
- Set all the symbology for the layers to coordinate each set of calls to their originating station, and make any necessary title changes or notes to make the map visually appealing.
- Save the results as **Exercise 5-9.mxd** in the \GIST2\MyExercises folder.

5-1
5-2
5-3
5-4
5-5
5-6
5-7
5-8
5-9

WHAT TO TURN IN

If you are working in a classroom setting with an instructor, you may be required to submit the maps you created in tutorial 5-9.

Turn in a printed map or screen capture of the following:

Tutorial 5-9.mxd

Exercise 5-9.mxd

Tutorial 5-9 review

Calculating the nearest facility along a network was far more accurate than the straight-line distance calculations. Natural barriers such as rivers, as well as areas with no roads, now play an important role in the results. For the purposes of the fire chief, this was a much more realistic and valuable method since the fire trucks are limited to driving on roads and don't handle cross-country driving very well.

Again, a good road network had to already exist, and it is important that it have good connectivity. If such a database didn't exist, it is possible to clean up an existing street centerline file to be suitable for network analysis.

STUDY QUESTIONS

1. What are the important differences between straight-line measurements and measurements along a network?
2. What are some of the requirements of a dataset to become a network database?
3. When would straight-line distance be a good model of reality?

Other real-world examples

A delivery company might want to use Network Analyst tools to determine which warehouse is the closest to a delivery site. Then the order could be dispatched using the most efficient route.

The public works department might build a network of the sewer lines to determine the nearest manhole to a reported problem. The distance may determine if the problem is reachable by a remote device from more than one access point.

The police department may want to analyze call data as the fire department did to determine if response times are in line with the distances the responding officers drove. A long response time to travel a short distance may point to problems in the dispatch system.

6

Mapping change

Temporal analysis deals with mapping change, which may occur as a change in location, a change in magnitude, or a change in one of the data's associated values. The change may be able to be shown on a single map, but may require a map series to show the results of the change. Features that change in more than one characteristic present an especially difficult challenge for cartographers. Carefully controlling the symbology and the amount of data being shown, as well as the number of maps used in a map series, will help make it easier to present the results.

Tutorial 6-1

Mapping change in location

Datasets capture values at one instant in time and paint a picture of that event. But some events happen over time, so data is collected to show how the values change during a specific time event.

Learning objectives

- *Combine datasets*
- *Track an event*
- *Overlay data*

Preparation

- *Read pages 149–164 in* The ESRI Guide to GIS Analysis, Volume 1.

Introduction

Much of the data you work with represents a single location at a single point in time. But data, like everything else, changes over time. So how can that be mapped?

Items can change in one of three ways: change in location, change in character, or change in value. These may be displayed on a series of maps in which the viewers must form pictures in their minds of each map and do some comparison between them. More visual analysis! Another method is to overlay the data onto a single map, using transparencies to make the data readable. These overlays can also be used to do other selections and quantify some result of the change.

When mapping the change over time for analysis, you may use the data readings to predict a future event. For instance, mapping the recorded path of a hurricane helps to predict where the storm may go next.

Mapping change over time can also be used to show conditions before and after an event. Data is recorded at one snapshot in time, then some event will take place, then more data is collected after the event to determine how things have changed. Recording accident information both before and after the installation of a new traffic control device might be used to show how effective the device is.

Mapping the time frame over which the change occurs is also a consideration in your analysis. The change may be recorded for specific dates or at a set interval. This could be an hour, a day, a week, a month, a year, and so on. The chosen time frame will depend on how fast the data changes. Mapping daily house values would not be very effective. Perhaps that is more of a monthly or annual thing. But mapping the expansion of a forest fire must be done within a short time frame.

When working with time-related data, you must also be aware of the duration, the number of values, and the interval. The duration represents the total time for which data is collected. The number of values is how many values are recorded over the duration, and the interval is the period between the time points when the values are recorded. An annual study with monthly values would have a duration of one year, twelve values, and an interval of one month.

You can combine the collected data with other data to see how other features, in addition to the recorded features, will change over time. Polygons showing the spread of a forest fire might be used to select the houses that have been consumed, giving a cost of destruction as the fire spreads.

Scenario A large producer of book matches is located in Oleander, and the company stores quite a bit of flammable material on-site for match production. The fire chief wants to do a "tabletop" disaster drill in which a fire breaks out at the match plant. You are asked to show the movement of the plume over the time frame of the drill.

Data The main dataset is a set of polygons created with the freely available ALOHA plume modeling software in order to predict the progression of a plume due to winds, temperature, and the chemical agent involved. There are four predictions, one hour apart.

Also available is a single layer with all of the plume data as well as a time-tracking field. This will be used for a Time Slider series.

The building footprint data is derived from aerial photographs, and each building includes a field called Use Code that represents the building's use. The codes are the following:

1	Residential	5	Government
2	Multi-Family	6	Utilities
3	Commercial	7	Schools
4	Industrial	8	Churches

Road and lot boundary layers are included for background interest.

6-1

6-2

6-3

Map the movement of a chemical plume

1   In ArcMap, open Tutorial 6-1.mxd. The match factory has had an accidental ignition, and the resulting fire has caused a release of potassium chloride into the surrounding neighborhood. The first hour's plume is shown, known as the Level 1 response plume. The factory's safety measures are containing the plume, but the fire chief has provided some data from a plume-tracking software that will show where the plume may travel over the next three hours. Use those predictive plumes to determine how many buildings must be evacuated.

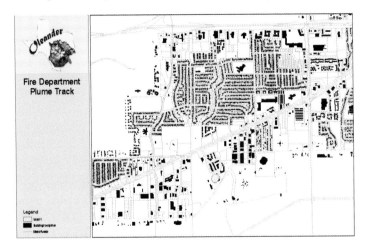

2   Open the properties of the Level 1 layer and go to the Symbology tab. Change the Fill Color to Mango.

3   Click the Display tab and change the transparency to 20%. Click OK to see the results in the map. Although you can do a very quick visual analysis and see how many buildings are affected, for a better understanding you should go through the process using GIS tools.

4   From the main menu, click Selection > Select By Location. Build the selection statement to select all the buildings within the plume. If the Building Footprints layer doesn't appear as an option, make sure the layer is selectable. When your dialog box matches the image, click OK.

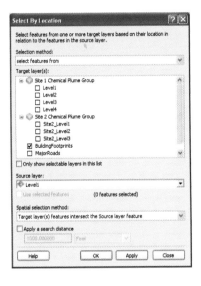

5 Open the attribute table for the
BuildingFootprints layer. Make a note of
how many buildings are selected and the
use code.

6 Close the attribute table. Clear the selected
features. Add a text box to the map's title bar to show how many buildings of each use
code are included in the Level 1 plume. A single building was selected. For the purposes
of the tabletop drill, the disaster area will be expanded and the response level elevated.

Note: Remember to use the navigation
tools on the Layout toolbar if you want to
zoom in for a closer look.

The ALOHA plume-modeling software
has predicted that the plume will expand
in one hour. This is represented by the
layer Level 2. Next try to determine what
structures, and their uses, will be affected
by this expanded plume.

7 Turn on Level 2. Open its properties and
change the color to Fire Red and set the
transparency to 50%.

6-1
6-2
6-3

8 Use the Select By Location tool to select the buildings that are inside the second plume.

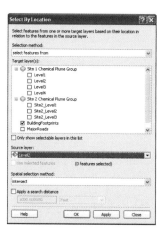

9 Perform a summary on the UseCode field to discover how many selected buildings belong to each use code. (**Hint:** See the image.) Add the results to the text box in the title block.

10 When you're finished, clear the selected features. The numbers you derived from this analysis will help the fire department plan for this scenario. The fire chief will be able to determine the resources needed over what time frame and decide how to deploy those resources.

YOUR TURN

Repeat the process with the Level 3 plume, making it Tuscan Red, and the Level 4 plume, making it Dark Umber. Determine the number of buildings in the plume, and calculate how many of each use there are. Add the results to the text box.

Set a transparency of 50 percent for all four plume layers. You can set this at the group level or individually. If you set them at the group level, make sure that there are no individual transparencies set, as these are additive.

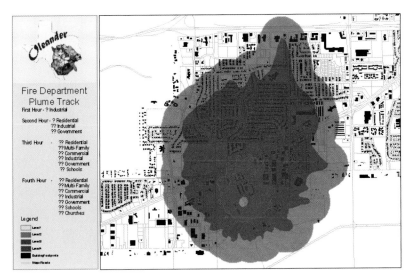

The exercise was a success. You learned how to map plumes and do some quick analysis; then the fire department was able to visualize the impact of these plumes over a four-hour duration.

Time Slider analysis

There is a new tool available to do time tracking, if you would like to have a visual representation of how the plume will grow over the four hours. The layer for the time series must contain all of the data and have a time field for each iteration that you wish to show. These are set up in the Time tab of the layer properties and then viewed with the Time Slider.

1 Turn off the Site1 Chemical Plume Group. Using the Catalog tab, navigate to the \GIST2\Data folder, and find the Fire Department feature dataset under the City of Oleander geodatabase. Drag the Site1_Time layer to the table of contents. The symbology is set up by time. Note that the event started at 5 AM and was measured once an hour for the next three hours. The Time Slider will be able to display the data according to the time it was collected.

2 Open the properties of the Site1_Time layer and go to the Time tab. In the Time window, you will need to identify the field containing the time information, its format, and the interval value.

3 Click Enable time on this layer. Set the Time Field to HourMeasur and Field Format to YYYY/MM/DD hh:mm:ss. Next, click Calculate to set the Layer Time Extent (duration). Set the Time Step Interval to 1 hour and click the box for Display data cumulatively. Click OK to close the dialog box.

With the duration and interval set up for the layer, you can view the data with the Time Slider window.

4 Click the Open Time Slider Window tool on the Tools toolbar.

5 In the Time Slider window, click the Enable time on map button; then click Play. The data will be shown in a time sequence. If the images go by too quickly, open the Time Slider options and set the speed slower. This is great for a visual representation but does not perform the analysis on building footprints you were able to do earlier. However, the flashy display is eye-catching. You can move the slider bar manually to show any specific time or just let the series play.

All of this was strictly visual analysis, using common ArcMap tools and ALOHA freeware. The value is in knowing how to string together these tools to achieve a different type of analysis, which was to show how something changed over time.

6 Save your map document as **Tutorial 6-1.mxd** in the \GIST2\MyExercises\ folder. If you are not continuing on to the exercise, exit ArcMap.

Taking it further

This tutorial used some real-life plume models for the match factory that had already been created. The process is a little long to have been included here, but you may be interested in learning to make your own plume models. The tutorial used the ALOHA plume-generating software, along with a custom script for importing the plume into ArcMap. These tools are free and can be downloaded from the NOAA Web site. To find them, go to `http://response.restoration.noaa.gov/` and search for Arc Tools.

Exercise 6-1

The tutorial showed the sequencing of a chemical plume and the process of selecting features inside the plume. In this exercise, you will repeat the process using another set of plume models.

- Continue using the map document you created in this tutorial or open Tutorial 6-1.mxd from the \GIST2\Maps folder.
- Move to the Site 2 Bookmark.
- Turn off the Site 1 Chemical Plume Group.
- Turn on the Site 2 Chemical Plume Group.
- Symbolize the layers as you did in the tutorial.
- Select the buildings that fall within each of the plumes.
- Calculate the number of buildings, and their use, that fall within each plume.
- Add an informative text box and change the titles, colors, and legend to make a visually pleasing map.
- Create a Time Slider series with the Site2_Time.lyr file.
- Save the results as **Exercise 6-1.mxd** in the \GIST2\MyExercises folder.

WHAT TO TURN IN

If you are working in a classroom setting with an instructor, you may be required to submit the maps you created in tutorial 6-1 and demonstrate the Time Slider series for the tutorial and exercise.

Turn in a printed map or screen capture of the following:

Tutorial 6-1.mxd

Exercise 6-1.mxd

6-1
6-2
6-3

Tutorial 6-1 review

This tutorial showed the change over time of a chemical plume. The duration was three hours, three datasets were given, and the interval was one hour. This was a predictive map, taking an event and predicting what would happen afterward.

Rather than do a time series of maps, you elected to overlay all the datasets and control how they are displayed using a transparency. For the fire department, the single display map worked better for distribution, whether on paper or a computer screen.

You also saw that once the datasets were made, you could use them for further analysis. They can be used to select other features or to perform other types of overlay functions. With the included time field, they could be used for a time-series analysis with the Time Slider tool.

STUDY QUESTIONS

1. How might you determine an interval value for a project?
2. What choices would you have for displaying the results of a time-series analysis?
3. How might one of the previous tutorials be designed as time-series analysis?

Other real-world examples

The public works department may map the flood level over time to predict street closing or evacuations.

The U.S. Census Bureau might use historical data to show population growth. This could then be used to plan transportation systems or distribute funding for various projects.

Soil patterns may be analyzed over time to determine erosion areas or the spreading of a river delta.

Tutorial 6-2

Mapping change in location and magnitude

Data that changes in location may also have a character or value associated with it that may also change. Through symbology, it is possible to map both the change in the physical location as well as show how the associated attribute values change.

Learning objectives

- *Symbolize values*
- *Understand variations in attribute values*
- *Create time-series maps*

Preparation

- *Read pages 165–167 in* The ESRI Guide to GIS Analysis, Volume 1.

Introduction

When values are associated with data that changes location, you have to look to see if the values are changing, too. In that instance, the data will have a change in both location and magnitude. The change in location can be symbolized with a linear path or by an expanding polygon as in the previous tutorial. The change in magnitude can be shown by changing the symbol's color or size.

You may also want to symbolize the path of the change with a line. This will not only be useful as showing the current track, but may be used as a predictive tool. It will also give the viewer an idea of how quickly the data is changing. If the line segments between locations are short, it shows that the location is changing slowly. Conversely, if the distance between locations is long, then the data values are changing rapidly, given that the study interval remains constant.

Scenario Last April, a tornado touched down in the northern part of the city. Through the local weather-gathering services, you were able to get the locations of the actual touchdowns, the tornado's magnitude, and the wind speeds as it moved through town. You want to map both the change in location as well as the changes in the tornado's strength and wind speed.

Data The tornado data was gathered through local observations. It includes the magnitude on the Fujita scale for tornado intensity and the wind speed in miles per hour.

6-1
6-2
6-3

The building footprint data is derived from aerial photographs, and each building includes a field called Layer that represents the building's use. The codes are the following:

1	Single Family
2	Multi-Family
3	Commercial
4	Industrial
5	Government
6	Utilities
7	Schools
8	Churches

Road and lot boundary layers are included for background interest.

Set graduated symbols

1 In ArcMap, open Tutorial 6-2.mxd. On April 14, a tornado ripped through town, tearing the roofs off many structures and leaving a general path of destruction. The city council needs a map of the damage path to secure relief funding. It needs to show the tornado as a dot proportional to its magnitude, the path as a linear symbol with variable width representing wind speed, and the damage buffer, which was determined from actual debris reports.

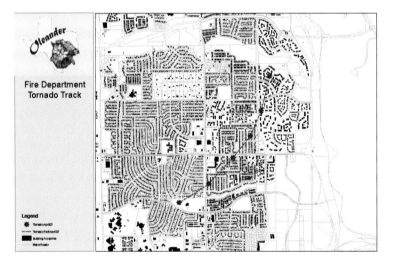

First you will set the symbology for the point layer of tornado sightings. This will indicate how strong the tornado was at each reading.

2 Open the properties of the TornadoApril07 layer. Set symbology type to Quantities > Graduated symbols with Fujita field as the Value. Select an appropriate color and dot size range, as well as the number of classes in the classification. Click OK.

Now you will symbolize the path of the tornado, showing a depiction of measured wind speed.

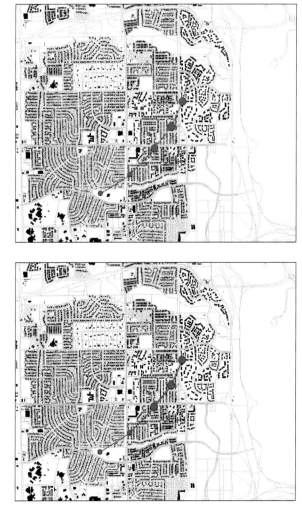

3 Open the properties of the TornadoPathApril07 layer. Change the symbology type to Quantities > Graduated symbols with the Value field of WindSpeed. Select an appropriate color and line-thickness range, as well as the number of ranges to display. Two things can be read in this visual display of the storm data. First, it is easy to see how the tornado increased and decreased in intensity over time. Second, it is possible to see the increase and decrease in wind speed over the same time period.

6-1
6-2
6-3

Buffer the tornado's path

Next, you'll add a buffer around the tornado's path that represents the damage area. Measurements were made and the distance was stored in an attribute called BufferDist for your use in this part of the analysis.

1 Search for and open the Buffer tool. Set the following parameters:

- Input Features: TornadoPathApril07
- Output Feature Class: TornadoPath_Buffer in the MyData.mdb geodatabase
- Distance Field: BufferDist
- Dissolve Type: All

When your dialog box matches the image, click OK.

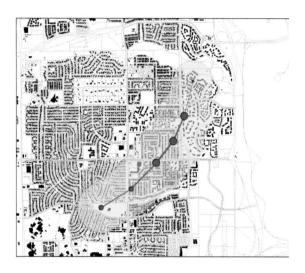

2 Open the Properties of TornadoPath_ Buffer. Set the color to Mango and the transparency to 50%.

Select affected buildings

Use this buffer area to select the affected structures. Use the Select By Location tool and the following description of the selection to be done to find the buildings that may have been impacted by the tornado:

> *I want to select the features from the layer BuildingFootprints that Intersect the features of the layer TornadoPath_Buffer.*

Remember that if the BuildingFootprints layer does not appear in the selection dialog box, it is because it has not been made selectable.

1 Start the Select By Location tool and fill in the parameters suggested in the above statement. When your dialog box matches the image, click OK.

Each step in the process is adding more valuable information. Now besides the visual display of the storm's intensity, it is possible to start seeing the area of damage the storm caused.

2 Open the attribute table for the BuildingFootprints layer and build a summary process on the UseCode field of the selected buildings. Save the table as **BuildingTotal** in the \GIST2\ MyExercises\ folder.

3 Clear the selection, then add a text box to the title block to show the numbers of buildings damaged. This map will now give the city council an idea of how many structures were damaged, as well as a visual representation of the wind speeds, tornado intensity, and ground speed of the tornado.

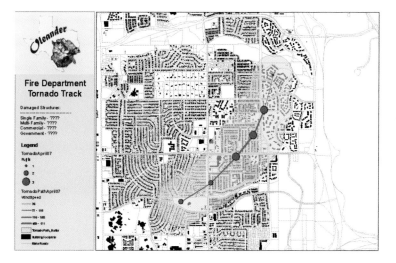

4 Save your map document as **Tutorial 6-2.mxd** in the \GIST2\MyExercises folder. If you are not continuing on to the exercise, exit ArcMap.

6-1

6-2

6-3

Exercise 6-2

The tutorial showed how you could display numerous aspects of a tornado with different symbology and be able to read damage area, ground speed, and wind speed of the tornado.

In this exercise, you will repeat the process using some data that the fire chief created for a drill scenario. You'll practice making the map, and the firefighters will evaluate the damage, evacuation plan, and other aspects of their planned response.

- Continue using the map document you created in this tutorial or open Tutorial 6-2.mxd from the \GIST2\Maps folder.
- Move to the Drill Area Bookmark.
- Turn off the Tornado Event Group layers.
- Turn on the Tornado Drill Group layers.
- Symbolize the Tornado touchdown locations and path as shown in the tutorial.
- Buffer the tornado path using the BufferDist field.
- Select the buildings that fall within the damage area.
- Calculate the numbers of buildings in each use category that fall within the damage area.
- Add an informative text box and change the titles, colors, and legend to make a visually pleasing map.
- Save the results as **Exercise 6-2.mxd** in the \GIST2\MyExercises folder.

WHAT TO TURN IN

If you are working in a classroom setting with an instructor, you may be required to submit the maps you created in tutorial 6-2.

Turn in a printed map or screen capture of the following:

Tutorial 6-2.mxd

Exercise 6-2.mxd

Tutorial 6-2 review

The observations of the tornado event were logged, along with the estimated intensity on the Fujita scale. This let you do a graduated symbol depiction of the storm intensity. You also worked with a linear event: the wind speed.

By showing this on the line, you were able to do a graduated symbol for the lines as well. The combination of thick lines and large dots highlighted the most intense periods of the storm.

The data can also be used for an overlay process. You buffered the lines a distance based on the recorded damage path and used this to select buildings. This could also have been overlaid on the parcel values to determine a quick damage estimate.

So through a combination of symbology on the lines and points, you were able to effectively communicate the changes both in location and magnitude.

STUDY QUESTIONS

1. How could you calculate the speed of an event based on the type of data you've been using?
2. How can magnitudes related to location data, both point and linear, be symbolized?
3. What technology might be used to collect point or linear data that demonstrates change in location?

Other real-world examples

The weather tracking stations might track hurricanes by noting the longitude, latitude, and intensity of each reading. This can be used for predictions before the fact or damage assessment after the fact.

The city engineering department might move a traffic-counting device along a busy street and take readings and then use the data to model traffic patterns at intersections.

A wildlife agency might track an animal herd, noting both the location and population through a migration pattern. A study could be done showing how the population changes as the animals move.

Tutorial 6-3

Mapping percent change in value

Data summarized by area, such as census counts or property values, changes over time. The change can be shown as simple values in a map series or as a percent change for the study interval.

Learning objectives

- *Understand database attributes*
- *Map percent change*
- *Perform time analysis*

Preparation

- *Read pages 168–173 in* The ESRI Guide to GIS Analysis, Volume 1.

Introduction

You may have worked with data that's collected for a single event and summarized for a given area. This may be the number of permits issued in a subdivision, the value of a piece of property, or the count of tigers in a wildlife reserve. For the one event, the numbers tell an important story.

When another round of data collection occurs, things get really interesting. Instead of just showing a single count event, you can now show how the values are changing over time: the change in the number of permits issued, the percent increase in property value, or the change in tiger density.

The difference in values can be shown simply as a change in the number, as a percent, or included in another calculation such as density. To get the change in the count, you just subtract the old value from the new value. This can be a positive or negative number, and might be symbolized with the negative values as red, the positive values as green, and the zero or no-change values as white.

Seeing the change for something like property value is usually pretty hard to comprehend over a large area, so it may be useful to show the change as a percent. To calculate this, you would subtract the old value from the new value, divide by the old value, and multiply by 100.

Finally, there are other means of showing the change, such as change in density. This is simply the difference of the values divided by the areas they represent.

After the change values are calculated, standard symbology techniques can be used to display the results.

Scenario The city council is concerned with property values in the city. It recently approved a new homeowner's assistance program that will help owners get grants to remodel their houses, thus increasing the values. It has also instituted a streamlined permitting process so that homeowners can have much of the work done quickly. You've been asked to make a map that will highlight the target areas—single-family residential property that has decreased in value more than 3 percent since 2004.

Data The parcel data includes a field for the appraised values in 2004 (TaxVal04) and a field for the appraised value in 2006 (TaxVal06). You'll use these to calculate a percent change.

Road and lot boundary layers are included for background interest.

Calculate percent change in property values

1 **In ArcMap, open Tutorial 6-3.mxd.** This map shows the parcels of Oleander with the symbology set to single symbol. You'll add a new field to the database, calculate the percent change in appraised value, and resymbolize the data.

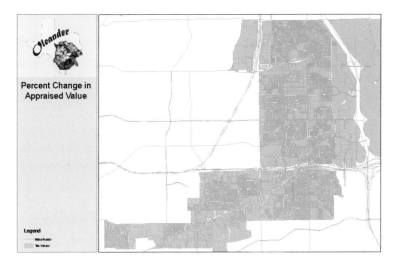

The program for funding passed by the city council is for residential property only, so step one will be to apply a definition query to the data to hide all the nonresidential property. You can do this very simply with the field DU that's been used before, which represents dwelling units. Only parcels with one dwelling unit qualify for this assistance.

2 Open the properties of the Tax Values layer and go to the Definition Query tab. Create the query **[DU] = 1** and close the layer properties.

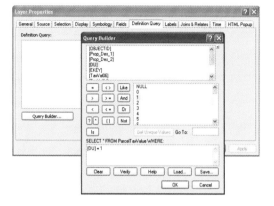

The next step will be to calculate the percent change in tax value from 2004 to 2006. You'll need a field to store the data, so you will add one to the attribute table. It will need to store decimal values, so its type will be floating point.

3 Open the attribute table for the Tax Values layer and select Table Options > Add Field. Name the new field **PercentChange06** and set the field type to Float.

4 Verify the new field has been added to the table, then close or dock the attribute table. There is a potential problem with the calculation for percent difference. In the 2006 data, there are parcels that have just been platted but have yet to be added to the county's tax rolls and consequently are valued at 0. You might also find some records in the 2004 data that might have been taken off the tax rolls or combined on a common ownership account.

Because of this, you'll also find values in that data with a value of 0. You will add an expression to the definition query to remove these so that the calculation will go smoothly. If this isn't done, it is likely that a "can't divide by zero" error will be generated. Will this new query be connected to the existing query with an AND or an OR? Write your answer before continuing.

5 Open the properties of the Tax Values layer, go to the Definition Query tab, and modify the query to exclude 0 values for TaxVal06 and TaxVal04 by adding **AND ([TaxVal06] <> 0 AND [TaxVal04] <> 0)**. When your dialog box matches the image, click OK, then OK again to close the properties. This will give you good values to use for the calculation.

Several parcels are not included in this study, but they represent the parcels that have changed ownership in the past two years and wouldn't qualify anyway. It's time to do the calculation. The formula for percent change in value is

$$(\text{New value} - \text{Old value})/\text{Old value} \times 100 = (\text{TaxVal06} \times \text{TaxVal04})/\text{TaxVal04} \times 100$$

6 Open the attribute table for the TaxValues layer. Right-click the new PercentChange06 field, select Field Calculator, and build the expression previously shown. (Read and dismiss the warning if you get one.) When your dialog box matches the image, click OK.

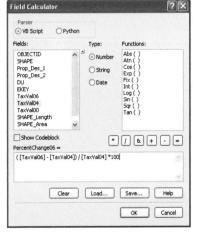

You'll notice that some values are 0, but this only means that their value didn't change. There are some negative values and some positive values. Now the new values can be symbolized.

Note: Once again, ArcMap is getting a sample of features on which to base the range of feature values. The default maximum may be reached. You can either ignore this error or change the sample size value by clicking the Sampling button on the Classification dialog box and increasing the maximum.

EKEY	TaxVal06	TaxVal04	TaxVal00	SHAPE_Length	SHAPE_Area	PercentChange06
1821725	172800	172800	145090	478.19408	13660.816906	0
2794322	189700	154400	127600	461.631642	13279.272794	22.862894
2794136	128400	118100	101500	460.291048	13217.009729	8.721422
1824392	160900	160900	132000	462.010137	13307.290991	0
6197299	130500	130500	97600	333.52503	6752.051127	0
1824430	164200	157200	130200	514.86242	16524.894763	4.452926
1823035	177700	177700	157300	649.427652	23268.704868	0
1822853	162500	162500	134860	530.464964	17575.460364	0

Symbolize results

The PercentChange09 field now holds a percent change value that can be used to symbolize the results. A graduated colors classification should make a good presentation. You may need to investigate how many classes to use. It would also be a good idea to use a color scheme that will show positive results in green and negative results in red.

1 Close the attribute table. Open the properties of the Tax Values layer and go to the Symbology tab. Change the Show method to Quantities > Graduated colors and set the Value field to PercentChange06. Use the Color Ramp drop-down menu to select a range from red to green. This looks pretty good, but the ranges need to be set manually to get just the segregation needed. You want to show houses that had no change as yellow, houses with a 0 percent to 3 percent change as light green, and houses with a greater than 3 percent change in dark green. At the other end of the scale, show a 0 to -3 percent change as orange, and a greater than -3 percent change as red.

2 Click the Classify button and set the Classification Method to Manual. Change Break Values to -3.001, -0.001, 0, 3, and 1600. When your dialog box matches the image, click OK.

3 Change the labels to better describe the values, as shown in the image. Right-click the top symbol and select Properties for All Symbols. Set Outline Width to 0. Click OK and OK.

4 Turn off the Lot Boundaries layer momentarily and look for areas with the highest negative change. Zoom to such an area. Turn the Lot Boundaries layer back on. The following image shows the area identified by the bookmark Target Zone. You may have chosen a different area. This resulting map will certainly give the city council an idea of where to concentrate their efforts to help stabilize the tax base. As it turns out, these houses are duplexes and might qualify for a federally funded project with matching money from the city. This would certainly help to leverage the local tax dollars to do more good.

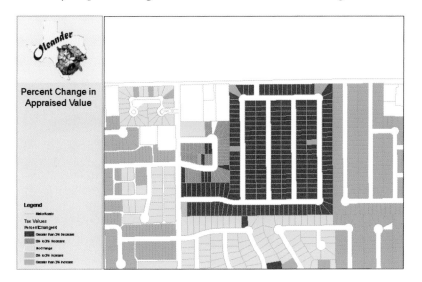

5 Save your map document as **Tutorial 6-3.mxd** in the \GIST2\MyExercises folder. If you are not continuing on to the exercise, exit ArcMap.

6-1

6-2

6-3

Exercise 6-3

The tutorial showed you how to calculate a change in a value from the attribute table, rather than a change in location.

In this exercise, you will prepare a map for the city council that shows the percent change in tax value from 2000 to 2004. Use the fields TaxVal00 and TaxVal04.

- Continue using the map document you created in this tutorial or open Tutorial 6-3.mxd from the \GIST2\Maps folder.
- Set the definition query for only parcels with 1 Dwelling Unit (DU).
- Add to the definition query a statement to exclude zero values.
- Add a field to the attribute table to contain the calculation.
- Perform the calculation for percent change.
- Set the classification manually to show no change, positive and negative changes of up to 3 percent, and positive and negative changes exceeding 3 percent.
- Zoom to an area that would be a likely candidate for the remodeling program.
- Change the titles, colors, and legend to make a visually pleasing map.
- Save the results as **Exercise 6-3.mxd** in the \GIST2\MyExercises folder.

WHAT TO TURN IN

If you are working in a classroom setting with an instructor, you may be required to submit the maps you created in tutorial 6-3.

Turn in a printed map or screen capture of the following:

Tutorial 6-3.mxd

Exercise 6-3.mxd

Tutorial 6-3 review

Changes you mapped in previous tutorials involved the feature moving, as well as an attribute changing. That was an easy one-to-one relationship with the location change and the value change. In this tutorial, the feature didn't move, but an attribute value changed. In order to show this, you needed a new attribute field that represented the change. If enough data was collected, it would be possible to create a time slider display as well.

This new value could represent only the numeric change or could be presented as a percent, like you did here. The change may have also been representative of some characteristic of the parcel, such as whether or not it has been platted in the past year. Instead of calculating a value, you could assign a new code to those parcels and use the code to set its symbology.

STUDY QUESTIONS

1. What role did time play in the tax data?
2. How might both the 2006 percent change from the tutorial and the 2004 percent change from the exercise be shown together?
3. Can you explain the unusually high percentage changes, like a positive 1600 percent increase?

Other real-world examples

The public library might want to map the change in the literacy rate for census tracts. Officials might look for areas with low literacy and initiate continuing education classes for reading. The effect of the reading campaign could be measured by showing new literacy levels over a period of time.

Census data is all about change from one dataset to another, with the coverage areas remaining the same. But even the Census Bureau deals with situations where the census blocks have zero values. These are areas like industrial parks or airports where there is no resident population. The inclusion of zeroes in the database can create problems calculating a change percentage.

The water district might track the stored water in a reservoir and show the percent change on a monthly basis. The polygons representing the lake shore wouldn't change over time, but the measured water level would. A map series could be created to show the water level measurement each month. The viewer would be able to understand how the levels have changed over the study period.

7

Measuring geographic distribution

Many of the analysis techniques covered so far deal with the relationships of features to other features, but there is another realm of analysis that deals with the characteristics of features within a single dataset. Measuring the distribution of your data—that is, the distances between features within a dataset—will begin to help you understand the possibility that your data may have patterns or clustering. Measuring the compactness of the data and applying statistical methods to the measurements can reveal otherwise unseen characteristics in the data.

Tutorial 7-1

Calculating centers

A distribution of features in a dataset has many characteristics that are helpful in performing analysis. One of these characteristics is the calculated center of the distribution. Centers can point out trends in movement, the best location to service all the features, or the feature that is the shortest total distance from the other features.

Learning objectives

- *Identify central feature*
- *Determine median center*
- *Determine mean center*

Preparation

- *Read pages 1–30 in* The ESRI Guide to GIS Analysis, Volume 2.

Introduction

Sets of features spread over some area are said to have a geographic distribution. Some might have a regular distribution, such as trees planted along a grid in an orchard. Other distributions might be seemingly random, such as locations of crime incidents. The goal of the spatial statistics tools in ArcGIS is to help identify and quantify spatial relationships and patterns. The results can then be used to compare datasets or track changes over time. The tools also help determine the statistical probability that the patterns and relationships really exist versus totally random occurrences. Fortunately, the spatial statistics tools in ArcMap make this type of analysis much simpler. Regular GIS consumers don't need to be doctorate-level statisticians to use the tools. You only need to understand how the tools function and when to apply them to get valid results.

The first measurement of geographic distribution is finding the center. There are three types of centers that are calculated for the data. The formulas for calculating these values can be rather complex and are illustrated on page 33 in *The ESRI Guide to GIS Analysis, Volume 2.*

The first type of center is the mean center. This is calculated by finding the average of the x-coordinate values of all the features, then finding the average of all the y-coordinate values. The resulting x,y coordinate pair is the mean center. This can be useful for determining the movement of a set of data over time, such as the migration of an animal herd.

The next type of center is the median center, also called the center of minimum distance. This function finds the x,y coordinate pair that is the closest to all the rest of the features. There is a built-in geographic weighting to this function, which does not occur with the mean center. This means that the median center will gravitate toward areas with a lot of features. This type of analysis might be used to find a location for a new facility that is closest to all the other facilities. The Median Center tool is not demonstrated in this tutorial, but more information is available about its use in *The ESRI Guide to GIS Analysis, Volume 2*.

And the third type of center is the central feature. This is the feature that has the lowest total distance to all the other features. Note that this must occur at a location that has a feature, unlike the median center, which can occur at any x,y coordinate location. The Central Feature tool also has a built-in geographic weighting, with groupings of features affecting which feature is selected. This could be used to find the school campus that is most accessible to all the others for a district-wide meeting.

Scenario The Fort Worth Fire Department in Tarrant County, Texas, has done some award-winning work with geospatial statistics. The fire chief, who is impressed with your previous maps, has agreed to let you work with the Fort Worth staff for this year's analysis and investigate the use of the spatial statistics tools. Not only will you be able to learn their techniques, but you can interject your own ideas and perhaps produce a better product.

Data The first dataset is the locations of the calls for service for February 2007. Each incident is located by both address and a coordinate pair for the incident in the Texas State Plane, North Central projection. These will have a built-in geographic weighting since areas with a lot of calls for service will tend to pull the centers in that direction.

The second set of data is the locations of the existing Fort Worth fire stations. Along with this data are the locations of some proposed stations. You might be able to look at some of the proposed stations and determine if the new locations would be better or worse for the recorded incidents.

Next are the Alarm Territories (AT). These territories are the primary response zones for each station. The field Station contains the number designation for each station. Several of these Alarm Territories are grouped together to form a Battalion Zone, which is monitored by a battalion chief. Any major incident occurring within an Alarm Territory will prompt the dispatching of the battalion chief for that Battalion Zone. The Battalion field in the attribute table contains the Battalion Zone number.

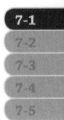

The fourth set of data is the street network data to give the map some context. It comes from the ESRI Data & Maps media.

Identify mean centers

1 In ArcMap, open Tutorial 7-1.mxd. The map displays the Battalion 2 area, which is in the central part of Fort Worth. The existing stations are shown, as well as proposed stations. You will find the mean center for all the calls and the central feature for Battalion 2.

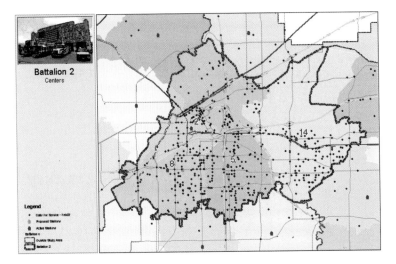

Battalion 2 contains several Alarm Territories (AT), and you want to calculate the mean center for each AT. The Mean Center tool will let you do this using the optional case parameter. You'll use the Station field for the case value, causing the tool to calculate a mean center for each unique value in the Station field. There are five ATs in Battalion 2, so you should get five mean centers. The dataset contains only the calls for service in Battalion 2 that originated from one of the Battalion 2 stations (Station 1, 2, 5, 8, or 14).

You'll run the Mean Center tool and compare the results with the station's location.

2 Use the Search window to locate and open the Mean Center tool.

3 In the Mean Center tool dialog box, set parameters as follows:

- **Input Feature Class: Calls For Service–Feb 07**
- **Output Feature Class: \GIST2\MyExercises\MyData.mdb\Bat_2_Mean_Centers**
- **Case Field: station**

When your dialog box matches the image, click OK.

4 Turn off the Calls For Service–Feb 07 layer. Change the symbology of the Bat_2_Mean_Centers layer to a large yellow circle. The result shows that most of the stations are located fairly close to their AT mean centers. Notice that the proposed new Station 8 is closer to the mean center than the existing station—this information might give the proposed location more credibility. It is important to note that when a fire unit from a station responds to a call outside of its AT, that location still has an effect on the mean center.

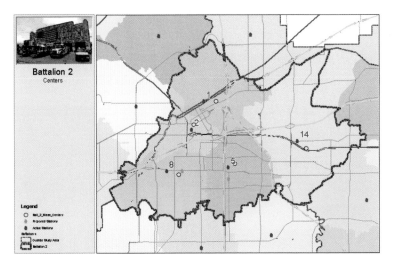

The Battalion 2 chief wants to relocate the battalion head office. Before the fire chief approves the money, it must be confirmed that the new location will be the most centrally located of all the stations in Battalion 2. This includes Stations 1, 2, 5, 8, and 14. Next you will select those stations and run the Central Feature tool.

Identify the central feature

Finding the central feature of Battalion 2 will first require that you select the active stations in that battalion. The tool will be run against the selected set.

1 From the main menu, click Selections > Select By Attributes. Build the following query to select the Battalion 2 stations from the Active Stations layer:

> **[STATION] = 1 OR [STATION] = 2 OR [STATION] = 5 OR [STATION] = 8 OR [STATION] = 14**

2 In the Search window, locate and open the Central Feature tool. An interesting option in the Central Feature tool is the distance method parameter. The Euclidean distance choice is the "as the crow flies" method, like the straight-line distances you worked with before. The Manhattan distance method

7-1
7-2
7-3
7-4
7-5

moves from the source to the destination using vertical and horizontal lines that turn at right angles. While this doesn't equate the distance along a network analysis that you did earlier, it is a better approximation in areas with a regular street grid.

3 In the Central Feature dialog box, set parameters as follows:

- **Input Feature Class: Active Stations**
- **Output Feature Class: \GIST2\MyExercises\MyData.mdb\Bat_2_Central_Feature**
- **Distance Method: Manhattan Distance**

4 When your dialog box matches the image, click OK.

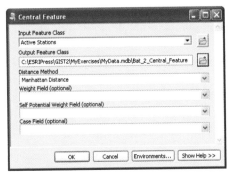

5 Set the symbology for the new layer to a big red star. Clear the selected features so that the symbology for all the layers will display correctly. The central feature is Station 2. Setting up the office at Station 2 will ensure that the battalion chief is located in the station most convenient to the other stations. Note that it doesn't mean that the chief will be any closer to expected calls for service.

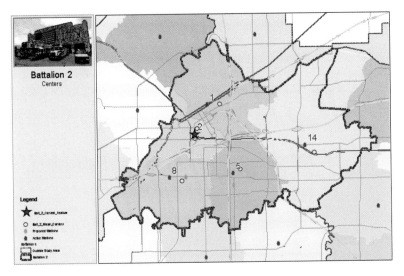

6 Save your map document as **Tutorial 7-1.mxd** in the \GIST2\MyExercises\ folder. If you are not continuing on to the exercise, exit ArcMap.

Exercise 7-1

The tutorial showed how to calculate a mean center and select a central feature.

In this exercise, you will repeat the process using different datasets. The Oleander city library has provided data about the patrons, and you want to locate three branch storefronts to serve them more conveniently. People would be able to go to the storefront locations to use the computers, participate in certain library programs, and reserve books that can be collected later.

Part 1   Find the mean center for each of the three districts.

- Open Tutorial 7-1E.mxd.
- Run the Mean Center tool for the Patron Locations layer. Set Case Field to A (which identifies the district).
- Symbolize the new layer appropriately.

Part 2   It is suggested that as part of the community outreach programs, the library should have on-site events at local apartment complexes. Using the Central Feature tool, find the best apartment complex in each district using the Apartment Complex polygons.

- Select District 1 in the Districts layer (Use the AreaName field).
- Select Apartment Complexes in District 1.
- Find the central feature of the Apartment Complexes.
- Repeat for Districts 2 and 3.
- Change the titles, colors, and legend to make a visually pleasing map.
- Save the results as **Exercise 7-1.mxd** in the \GIST2\MyExercises folder.

WHAT TO TURN IN

If you are working in a classroom setting with an instructor, you may be required to submit the maps you created in tutorial 7-1.

Turn in a printed map or screen capture of the following:

        Tutorial 7-1.mxd

        Exercise 7-1.mxd

7-1
7-2
7-3
7-4
7-5

Tutorial 7-1 review

Finding the centers of a set of features provided some interesting data for analysis. The mean center of the call for service data gave a good idea of where a fire station might best be located. It would be interesting to calculate the mean center monthly and see how it moves throughout the year. With data over the time span of one year, it may become a useful predictive tool for winter calls versus summer calls.

Finding the central feature allowed the battalion chief to find a new office that might minimize travel time. The total distance from each station to every other station was calculated, with the lowest total being chosen. Repeating this operation in the library exercise demonstrated that this function works on polygons as well.

It is important to remember that these two processes had a geographic weighting built in. Groupings of features tended to draw the results closer to them than areas with few or widely spaced features.

STUDY QUESTIONS

1. Would it be valuable to find the central feature of calls for service? Why or why not?
2. What other factors might be considered in locating a facility besides using the mean center of customers?

Other real-world examples

A police department may map crime locations by type and calculate the mean center to determine where to focus a task force.

A company may look for a new office location based on the mean center of their customers or employees.

A banking company might schedule regional meetings at the branch calculated as the central feature to reduce overall travel time for employees.

Tutorial 7-2

Adding weights to centers

Considering all features equally when calculating centers may not give the most realistic results when other factors may be affecting the distribution of the values. Adding weights to these calculations will make the centers gravitate toward features with more importance to the analysis.

Learning objectives

- *Identify weighted central feature*
- *Determine weighted median center*
- *Determine weighted mean center*

Preparation

- *Read pages 30–38 in* The ESRI Guide to GIS Analysis, Volume 2.

Introduction

Some features in a geographically distributed set may carry more significance, or weight, than others. An option in the functions to calculate centers lets you use these weights to draw the centers closer to these features. For example, using the locations of businesses to calculate a mean center might put the location in one place. But factoring in the number of employees at each business might pull the mean center toward the businesses with more employees.

In order to add a weight to a center calculation, the features need to have an attribute in their table that expresses the weight as a number. This may be a count or a code, provided that it is a number expressing the feature's importance.

Scenario   You are once again working with the Fort Worth Fire Department. It has provided a value for incident type that ranks its importance on a scale of 1 to 10. It seems that some of the calls logged in the database, such as false alarms and stand-by actions, were factored into the mean centers with equal weight. If the mean center method is to be used to evaluate the possible locations of fire stations, the chief would like to give higher priority calls more significance in the calculations. The highest-priority calls are those that involve the preservation of life, with medium-priority calls dealing with the protection of property. The lowest-priority calls are false alarms and minor property destruction such as trash-bin fires. You will recalculate the mean center using the weights provided.

Data The Calls For Service–Feb 07 database has a field called FEE (Fire/EMS Evaluation). This field has values from 1 to 10, with 10 representing the most important calls.

In addition, the Active Stations layer has a field called NumEmployees that represents the number of employees at each station.

Identify weighted mean centers

1 In ArcMap, open Tutorial 7-2.mxd. The map display looks just like the previous tutorial. You will do the same process to calculate the centers, except this time use the FEE field as the weight for the mean center, and the NumEmployees field as the weight for the central feature.

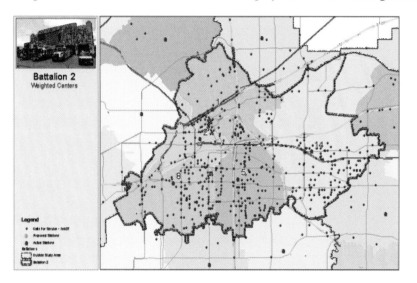

The data has already been restricted to only the calls for service from Stations 1, 2, 5, 8, and 14.

2 Open the attribute table for the Calls For Service–Feb 07 layer. Scroll through the records, noting the "FEE" and "descript" fields to get an idea of the incident types that get the highest weight. Close the table when you are done.

3 Locate and run the Mean Center tool and set parameters as follows:

- **Input Feature Class: Calls For Service–Feb 07**
- **Output Feature Class:**
    \GIST2\MyExercises\MyData.mdb\Bat_2_Weighted_Mean_Centers
- **Weight Field: FEE**
- **Case Field: station**

4 When your dialog box matches the image, click OK.

5 Turn off the Calls For Service–Feb 07 layer and make the symbol for the new layer a large green circle. Add the layer Bat_2_Mean_Centers, which you created in Tutorial 7-1 and symbolize as a large yellow circle. The weights have moved the mean center for each Alarm Territory slightly. In this case, the effect of the weights was minimal, but some weights may alter the results dramatically.

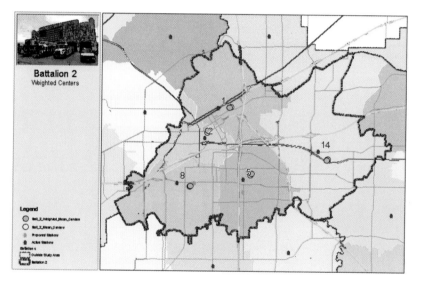

Next you will try the Central Feature command using the NumEmployees field as a weight. The chief would like to be at the location that gives the best access to the firefighters. Remember that in the previous tutorial, Station 2 was selected as the new home for the battalion chief.

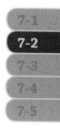

Identify the weighted central feature

1 Use the Select By Attributes tool to select only the stations in Battalion 2 (1, 2, 5, 8, and 14).

2 Locate and open the Central Feature tool and set parameters as follows:

- **Input Feature Class: Active Stations**
- **Output Feature Class:**
  \GIST2\MyExercises\MyData.mdb\Bat_2_Weighted_Central_Feature
- **Distance Method: Manhattan Distance**
- **Weight Field: NumEmployees**

3 When your dialog box matches the image, click OK. With some features having more significance in the calculation, or a larger weight value, you would expect to see a shift in the location of the central feature.

4 Change the symbology of the new central feature layer to a large green star.

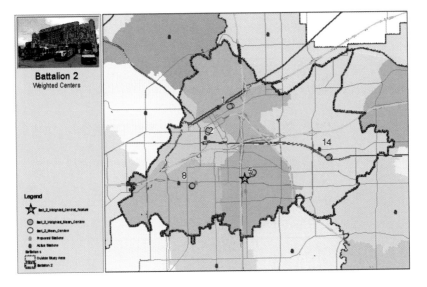

With the added importance of the employee count, the new battalion chief location has moved from Station 2 to Station 5.

5 Save your map document as **Tutorial 7-2.mxd** in the \GIST2\MyExercises folder. If you are not continuing on to the exercise, exit ArcMap.

Exercise 7-2

The tutorial showed how to add an optional weight value, which gave some features more significance in the calculation to find the center.

In this exercise, you will repeat the library analysis from the exercise in the previous tutorial, adding a weight value. The library database was summed for each address, so the layer Patron Locations contains a field that summed the number of visits to the library from each address. Some of the addresses are apartment complexes, so the number for these addresses gets quite high.

Part 1   Find the weighted mean center for each of the three districts. The field A in the table for Patron Locations can be used as the case field, and the field Usage can be used as the weight field.

- Open Tutorial 7-2E.mxd.
- Run the Mean Center tool for the Patron Locations layer. Set the Weight Field to Usage and the Case Field to A.
- Symbolize the output appropriately.

Part 2   The Apartment Complex layer contains a field called DU, which represents the number of dwelling units at each site. Use this as the weight field for the central location calculation so it will take into account the number of units, not just the geographic location.

- Select District 1 in the Districts layer.
- Select apartment complexes in District 1.
- Find the central feature of the apartment complexes, with the weight field set to DU.
- Repeat for Districts 2 and 3.
- Change the titles, colors, and legend to make a visually pleasing map.
- Save the results as **Exercise 7-2.mxd** in the \GIST2\MyExercises folder.

WHAT TO TURN IN

If you are working in a classroom setting with an instructor, you may be required to submit the maps you created in tutorial 7-2.

Turn in a printed map or screen capture of the following:

Tutorial 7-2.mxd

Exercise 7-2.mxd

7-1
7-2
7-3
7-4
7-5

Tutorial 7-2 review

Adding a weight to the mean center calculations lets you see how another characteristic of the features could affect the results. The features have a built-in geographic weight, but other attributes such as a value or count can further enhance the reality of the calculation.

The central feature calculation was also affected by the inclusion of a weight. By adding another characteristic to the calculation, the result was drawn toward the feature with the largest weight value.

STUDY QUESTIONS

1. What types of values might be useful as a weight value? Which types might not be?
2. What is the purpose of adding a weight to the calculations?

Other real-world examples

A police department may weight crime locations by severity to assign a specific task force to the areas with the most dangerous crimes.

A company planning to open a new office may weight the calculation of a central location by the number of transactions attributed to each customer in order to select a site nearest the best customers. This would reduce the average travel time for customers.

A banking company looking for a site to hold regional meetings might weight the calculation of a central location by the number of employees at each branch to reduce the overall driving distance for all employees.

Tutorial 7-3

Calculating standard distance

Measuring the compactness of a distribution provides a single value representing the dispersion of features around the center. This value is calculated by measuring the average distance of the features from the mean center and how much the individual distances vary from the average distance. This is called the standard distance deviation, or simply the standard distance.

Learning objectives

- *Determine compactness of distribution*
- *Work with weight values*

Preparation

- *Read pages 39–44 in* The ESRI Guide to GIS Analysis, Volume 2.

Introduction

Measuring the standard distance lets you quantify the amount of dispersion in a set of features. It is calculated by determining the average distance each feature is from the mean center, then determining a value for how much the distances deviate from the average distance. The results are shown on a map as circular rings of one, two, or three standard deviations. As with the calculation of mean centers, a natural geographic component of each feature is used in calculating standard distance.

Standard distance circles can also be given a weight, similar to the weighted centers. Besides the natural geographic components, an additional factor can be introduced that may influence the results. If a weight is given, then the results are measured around the weighted mean center.

It is important to note that a standard distance circle does not determine grouping or clustering, only that the features are occurring in close proximity. A standard distance circle may include several groupings or clusters. These topics will be discussed later.

Scenario   You will continue the analysis of the Battalion 2 data for the Fort Worth Fire Department. The goal is to show the standard distance for each station. You want to calculate the standard circle both with and without the weight value FEE, which will add more importance to some features than others. From this you will be able to determine which station's units are making more runs outside of their normal Alarm Territory.

7-1
7-2
7-3
7-4
7-5

Data   The Calls For Service–Feb 07 database has a field called FEE (Fire/EMS Evaluation). This field has values from 1 to 10, with 10 representing the most important calls.

The last set of data is the street network data to give the map some reference. It comes from the ESRI Data & Maps media.

Determine standard distance

1   In ArcMap, open Tutorial 7-3.mxd. The five stations from Battalion 2 will be used for the standard distance calculations. You will compare across districts, which may highlight stations that are understaffed.

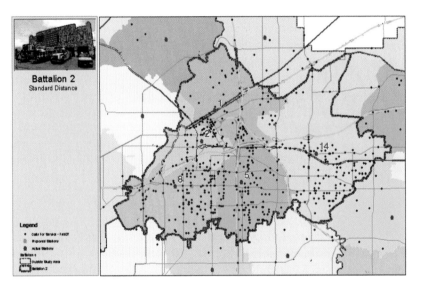

2   Locate and open the Standard Distance tool with the Search window.

3   In the Standard Distance dialog box, set parameters as follows:

- **Input Feature Class: Calls For Service–Feb 07**
- **Output Feature Class: \GIST2\MyExercises\MyData.mdb\Bat_2_Stnd_Dist**
- **Case Field: station**

4   When your dialog box matches the image, click OK.

The results will be five circles, one standard deviation from the mean center of each Alarm Territory.

5 Set the color of the new layer to Fire Red and set transparency to 60%. You can see that stations 5 and 8 have a fairly compact standard distance circle. Their average travel time to incidents is mostly within the boundaries of their alarm response territory. Stations 1 and 14 have large standard distance circles, which show that their average travel time is relatively long.

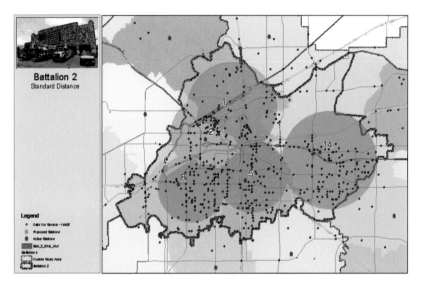

These circles do not take the relative importance of the incidents into account. You are more concerned with the average travel time to the most important calls. So if these are adding to the size of the standard distance, you might want to consider adding another station to lower the average response time. Run the process again using the field FEE as the weight.

Determine standard distance using weights

1 Open the Standard Distance tool and set parameters as follows:

- **Input Feature Class: Calls For Service–Feb 07**
- **Output Feature Class:**
        \GIST2\MyExercises\MyData.mdb\Bat_2_Weighted_Stnd_Dist
- **Weight Field: FEE**
- **Case Field: station**

2 When your dialog box matches the graphic, click OK.

7-1

7-2

7-3

7-4

7-5

3 Set the color for the new layer to Yellow with a Transparency of 60%. Select all the features in this layer by right-clicking the layer in the table of contents and clicking Selection > Select All.

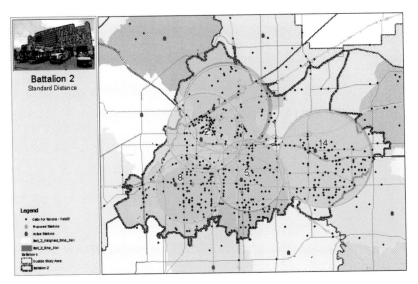

Compare the weighted and unweighted standard distance circles. Stations 5 and 8 did the best with the unweighted standard distance. They also fared well with a positive change for Station 5, indicated by the reduction in size of the weighted standard distance circle when compared to the unweighted standard distance circle. Of the two stations that did the worst before, Station 1 shows the greatest improvement in reducing the average time. So when you add an incident priority code to the analysis, the performance of the stations looks better.

4 Save your map document as **Tutorial 7-3.mxd** in the \GIST2\MyExercises folder. If you are not continuing on to the exercise, exit ArcMap.

Exercise 7-3

The tutorial showed how to calculate both weighted and unweighted standard circles in order to assess the compactness of feature distribution.

In this exercise, you will repeat the process using the data from the library again. You will be computing the standard distance for each district, then repeating it with a weight. The weight field will be Usage, which represents the number of library transactions attributed to each feature. Note that some usage values seem unusually high because they are summed for apartment complexes.

Part 1   Find the standard distance for each of the three districts. The field A in the table for Patron Locations can be used as the Case field.

- Open Tutorial 7-3E.mxd.
- Run the Standard Distance tool. Set Case Field to A.

Part 2   Run the Standard Distance tool again with a weight. The field A is again used as the Case Field, and the Weight Field is Usage.

- Change the titles, colors, and legend to make a visually pleasing map.
- Save the results as **Exercise 7-3.mxd** in the \GIST2\MyExercises folder.

WHAT TO TURN IN

If you are working in a classroom setting with an instructor, you may be required to submit the maps you created in tutorial 7-3.

Turn in a printed map or screen capture of the following:

> **Tutorial 7-3.mxd**
> **Exercise 7-3.mxd**

7-1
7-2
7-3
7-4
7-5

Tutorial 7-3 review

You created both weighted and unweighted standard distance circles in this tutorial. It was apparent that many of the features used for standard distance fell within the circle, while other features fell outside the circle. The features outside are affecting the size of the circle, and thus the compactness of the distribution. With the weight field applied, the features outside the circle may not have affected the calculations in a negative way if their weight value was low. In most cases you saw that adding a weight reduced the size of the circle.

It is important to note that features that may have belonged to different geographic clusters still had a single circle to represent the distribution. Once again, the standard distance circles show compactness of the overall distribution, but do not show groupings or clusters.

STUDY QUESTIONS

1. What does the Standard Distance tool measure?
2. What is the significance of the weight when calculating a standard distance?
3. What would the data-distribution chart look like for data with a small standard distance? A large one?

Other real-world examples

A police department may measure the compactness of a set of crime data to determine a possible patrol area. A concentrated area would be easy to patrol with extra officers, while a dispersed area may not highlight a clear tactic.

A county health service may look at the standard distance for West Nile Virus cases to determine a mosquito-treatment area. A small standard circle may highlight an area in which to focus treatment.

Tutorial 7-4

Calculating a standard deviational ellipse

Compactness is only one characteristic of distribution that may be of interest. Another is directional trend. The standard deviation is calculated separately for the x- and y-axes, resulting in an elongated circle, or an ellipse.

Learning objectives

- *Identify directional trends*
- *Work with weight values*

Preparation

- *Read pages 45–50 in* The ESRI Guide to GIS Analysis, Volume 2.

Introduction

The distribution of features may have a directional trend, making it hard to represent with a circle. By calculating the variances for both the x- and y-axes independently, the directional trend can be shown. The result is shown with an ellipse, with the axes oriented to match the orientation of the features.

The ellipse is referred to as the standard deviational ellipse since the method calculates the standard deviations of the x-coordinates and the y-coordinates from the mean center. When the distribution is elongated along an axis, the ellipse will reflect the orientation of that axis. While visual analysis may give you an idea of the orientation of the features, the standard deviational ellipse may give you more confidence in your results since it is based on a statistical calculation.

With the addition of an orientation component, standard deviational ellipses can more accurately depict the distribution of data than the standard distance. Features that group along a path would create a large standard distance, or a thin, elongated standard deviational ellipse. This may reveal a trend along a road or stream, for instance.

As with the standard distance, you can calculate the standard deviational ellipse either using the locations of the features as a built-in geographic weight, or allowing the locations to be influenced by a particular attribute value. The latter would be the weighted standard deviational ellipse.

7-1
7-2
7-3
7-4
7-5

Scenario You are going to revisit the Battalion 2 data for the Fort Worth Fire Department. The freeways going through the region may give more of a directional trend to the data that can be revealed with the standard deviational ellipse. You want to calculate the ellipse both with and without a weight based on the FEE attribute, which will add more importance to some features than others. From this you will be able to visualize any directional influence that the roads may be having on the data.

Data The Calls For Service–Feb 07 database has a field called FEE (Fire/EMS Evaluation). This field has values from 1 to 10, with 10 representing the most important calls.

The map also includes street network data to give the map some reference. It comes from the ESRI Data & Maps media.

Use the Directional Distribution tool

1 In ArcMap, open Tutorial 7-4.mxd. Once again you get the familiar Battalion 2 area. These five stations will be used for the standard deviational ellipse calculations. Comparing the ellipses across districts may give you an idea of any directional trends.

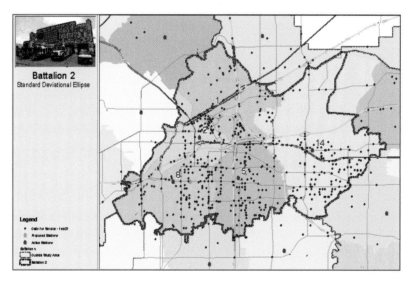

2 Use the Search window to locate and open the Directional Distribution tool.

3 In the Directional Distribution dialog box, set parameters as follows:

- **Input Feature Class: Calls For Service–Feb 07**
- **Output Ellipse Feature Class:**
     \GIST2\MyExercises\MyData.mdb\Feb07StndDevEllipse
- **Case Field: station**

4 When your dialog box matches the image, click OK. The standard deviational ellipse will be calculated and added to the table of contents, with one for each occurrence of the case field. Note the shape and orientation of the ellipses, and match the directional trend to both the roads and the features.

5 Open the properties of the Feb07StndDevEllipse layer and set a transparency value of 40%. You can easily see that the calls from Station 5 have a definite north–south trend, following the freeway. Station 8 has an east–west trend, following another freeway. Station 2's orientation is affected by calls in a particular neighborhood, and not oriented to the roads, while Station 14's broad response area and even distribution keep its ellipse very close to being a circle.

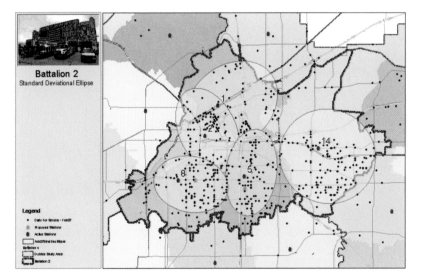

Next you will try this again using the rankings in the FEE field as a weight.

Use the Directional Distribution tool with weights

1 Open the Directional Distribution tool again and set parameters as follows:

- **Input Feature Class: Calls For Service–Feb 07**
- **Output Ellipse Feature Class:**
  \GIST2\MyExercises\MyData.mdb\Feb07StndDevEllipseWght
- **Weight Field: FEE**
- **Case Field: station**

7-1
7-2
7-3
7-4
7-5

2 When your dialog box matches the image, click OK.

3 Open the properties of the Feb07StndDevEllipseWght layer and set transparency to 40%. Notice that the orientation of Station 5's ellipse is narrower with the weights applied, and the distribution of Station 14 calls is slightly less circular. This is only one month of data, and you would want to look at the change in the ellipses over a longer time to determine the most optimal station placement.

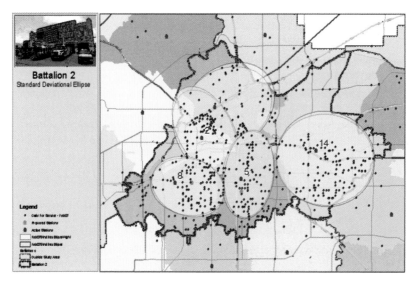

4 Save your map document as **Tutorial 7-4.mxd** in the \GIST2\MyExercises folder. If you are not continuing on to the exercise, exit ArcMap.

Exercise 7-4

The tutorial showed how to calculate both weighted and unweighted standard deviational ellipses. This not only showed compactness of the data, but highlighted directional trends.

In this exercise, you will repeat the process using the data from the library again. You will be computing the standard deviational ellipse for each district, then repeating it with a weight. The weight field will be Usage, which represents the number of library transactions attributed to each feature. Note that some usage values seem unusually high because they are summed for apartment complexes.

Part 1   Find the standard deviational ellipse for each of the three districts. The field A in the table for Patron Locations can be used as the case field.

- Open Tutorial 7-4E.mxd.
- Run the Directional Distribution tool. Set Case Field to A.

Part 2   Run the Directional Distribution tool again with a weight. The field A in the table for Patron Locations is again used as the case field, and the weight field is Usage.

Part 3   Add the weighted standard distance circles from exercise 7-3 and compare the results.

- Change the titles, colors, and legend to make a visually pleasing map.
- Save the results as **Exercise 7-4.mxd** in the \GIST2\MyExercises folder.

WHAT TO TURN IN

If you are working in a classroom setting with an instructor, you may be required to submit the maps you created in tutorial 7-4.

Turn in a printed map or screen capture of the following:

Tutorial 7-4.mxd

Exercise 7-4.mxd

7-1
7-2
7-3
7-4
7-5

Tutorial 7-4 review

You created both weighted and unweighted standard deviational ellipses. In addition to showing data compactness, like the standard distance circle, the ellipse can show a directional trend. The weights for the fire response data tended to draw the ellipse toward the freeways, where a large number of traffic accidents may have had a large effect.

If the ellipses remain more circular, it's an indication that the data does not have a directional trend. The data shows that calls from Station 14 were spread all across the district. If many of the calls for service all around the edges of the district have a large weight, the ellipse will not appear much different than the standard distance circle. This would indicate that the data did not have much directional trend.

STUDY QUESTIONS

1. What information do the ellipses provide that the circles do not?
2. What is the relationship between the weighted center and the weighted standard deviational ellipse?
3. What is the difference between the weighted standard deviational ellipse and the weighted standard circle?

Other real-world examples

A police department may look at the standard deviational ellipse to spot directional trends in crime, such as following a hiking trail or freeway.

Animal-tracking data may be mapped with a standard deviational ellipse to see directional trends in migration or grazing.

Tutorial 7-5

Calculating the linear directional mean

The directional trend of linear objects is calculated using a different methodology than points or polygons. The lines are analyzed to determine their angles, then the mean angle is calculated. The result will demonstrate a directional mean as an arrow.

Learning objectives

- *Identify directional trends of linear features*
- *Work with case values*

Preparation

- *Read pages 51–61 in* The ESRI Guide to GIS Analysis, Volume 2.

Introduction

Calculating the linear directional mean can result in one of two outputs, depending on the input data. These are the mean direction and the mean orientation. Each is based on the average angle of the lines in the dataset, but mean direction takes into account the direction of movement.

When lines are created in ArcMap, they have a starting point and an ending point. The deviation of a line from the horizontal is measured as an angle from the starting point to the ending point. Then each of the lines is transposed so that their starting points are coincident and a mean angle is calculated. The mean direction is useful for data that represents features that move, like hurricanes or wildlife. The results of the Linear Directional Mean tool are then displayed with an arrow to show direction.

If the line features represent stationary items, such as fault lines or freeway systems, they do not have a direction. Instead, the line created by the Linear Directional Mean tool shows only orientation. This might be shown with a double-headed arrow.

Before using this tool, be sure to understand which result you will be getting. If the mean direction is desired, be sure to make all the line features point in the right direction with respect to their starting and ending points. Another thing to note is that for multisegment lines, only the starting and ending points of the feature are used to calculate that feature's angle. Vertices are not used in the calculation. So an "S" shaped feature would be represented in the equation by a single line from the start point to the end point.

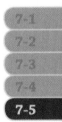

Scenario　As part of its disaster planning, the Tarrant County Emergency Operations Center would like to analyze past tornado events in the county. After some research, archival data was assembled into a GIS format. You are asked to find the linear directional mean of each year's storms in order to better predict future occurrences.

Data　The layer Tornado has single point measurements of tornado touchdowns. The attribute table includes the year, latitude and longitude coordinate pair, and the tornado's measured intensity on the Fujita scale.

The layer Storm Track contains lines that connect the tornado touchdown points for each individual storm. Since the lines were digitized by connecting the first touchdown point of each tornado with its last, they will not only produce an angular trend, but also a directional trend.

The other layers are the Census 2004 population and road network for background interest.

Determine the linear directional mean of tornadoes

1　In ArcMap, open Tutorial 7-5.mxd. The map shows the 2004 population in purple, along with a street network. The storm tracks are the paths each year's recorded tornadoes took after touchdown. You will calculate the linear directional mean to see what the directional trend was for each year's storms. This will give the fire department an idea of where future tornadoes may go after touchdown.

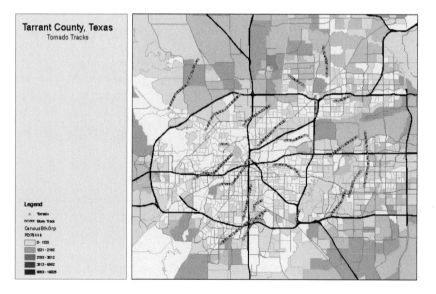

2　Use the Search window to locate and open the Linear Directional Mean tool.

3 In the Linear Directional Mean dialog box, set parameters as follows:

- **Input Feature Class: Storm Track**
- **Output Feature Class: \GIST2\MyExercises\MyData.mdb\StormLinearDirMean**
- **Case Field: YEAR_**

4 When your dialog box matches the image, click OK. The results are lines representing the angular means of the sets of storms for each year. To show direction, the symbology has been changed to an arrow.

5 Open the properties of the new StormLinearDirMean layer and go to the Symbology tab. Click the symbol to open the symbol selector, and set the symbol to Arrow at End, Mars Red with a width of 5. (See if you can figure out how to make the arrowhead red as well.)

6 Next, go to the Labels tab on the properties and set Label Field to YEAR, font size to 14, and color to red. Check the box to turn the labels on and exit the properties dialog box.

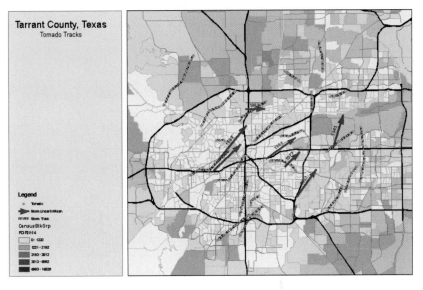

The storms seem to move in the northeast direction. To see the trend for a specific year, select only the 1977 storms.

7 On the main menu, click Selection > Select By Attributes. Using the Storm Track layer, build the query **[YEAR_]** = '**1977**'. When your dialog box matches the image, click OK.

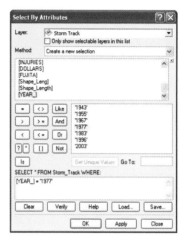

You can see the four storm tracks used to make the 1977 directional mean line. The directional mean was calculated, and a line drawn at the mean angle was added at the mean center of the selected storm tracks. The length of the directional mean line is the average length of the storm track lines.

8 Save your map document as **Tutorial 7-5.mxd** in the \GIST2\MyExercises folder. If you are not continuing on to the exercise, exit ArcMap.

YOUR TURN

Select the storm tracks for the other years to see their trends. Open the SelectBy Attributes dialog box and build similar queries for 1943, 1955, 1967, 1983, 1996, and 2003. Compare the lengths of the selected storm track lines with the length of the calculated linear directional mean for each set.

Exercise 7-5

The tutorial showed how to calculate the linear directional mean for the tornado tracks in Tarrant County. This gave you the directional trends of the storms and the mean path length of each year's storms.

In this exercise, you will repeat the process using the data from Dallas County. The storm tracks have been mapped, with the field YEAR showing in which year each storm occurred.

- Open Tutorial 7-5E.mxd.
- Run the Linear Directional Mean tool. Set the Case Field to YEAR.
- Compare the path lengths of each year's set of storms with the length of that year's linear directional mean.
- Change the titles, colors, and legend to make a visually pleasing map.
- Save the results as **Exercise 7-5.mxd** in the \GIST2\MyExercises folder.

WHAT TO TURN IN

If you are working in a classroom setting with an instructor, you may be required to submit the maps you created in tutorial 7-5.

Turn in a printed map or screen capture of the following:

Tutorial 7-5.mxd

Exercise 7-5.mxd

7-1
7-2
7-3
7-4
7-5

Tutorial 7-5 review

The Linear Directional Mean tool created single lines to represent the directional trend of the tornadoes in the study area. Since the storm tracks had a direction, and the lines were verified to have been digitized in the correct order, the results can be symbolized with an arrow.

The length of the output line represents the mean length of the storm tracks. Keep in mind, though, that path length may not be indicative of the damage the storm caused.

STUDY QUESTIONS

1. Which would produce the longest output vector: a lot of short storm tracks or a few long ones? Why?
2. What planning might the fire department do for an approaching storm if it knew the directional mean of storms from the past 10 years?
3. How does mean orientation differ from mean direction?

Other real-world examples

A police department can map the track from the point a car is stolen to the point of recovery. This gives a directional mean that can help predict the actions of an auto theft ring.

Wildlife biologists use animal-tracking data to map migration paths. The linear directional mean of these paths may give some insight into why certain paths are chosen or highlight the need for protected habitat.

Transportation studies often match traffic counts with the mean orientations of freeway systems. The result may show where there are too few roads to handle the traffic and may help plan the system's expansion.

8

Analyzing patterns

The statistical analysis tools used to study patterns in a dataset can answer a global question: "What is the probability that the distribution of these features is occurring due to random chance?" These tools don't create anything that can be symbolized on a map, although it may be helpful to capture some of the graphics that they create. The tools are important, however, in establishing a statistical foundation and a statistical confidence level for pattern analysis.

Tutorial 8-1

Using average nearest neighbor

When working with spatial statistics to identify patterns, various tools are used to measure a characteristic of the data that indicates if the data is thought to be clustered or dispersed; or if in fact it occurs randomly. This measure, or index, is then tested to see if statistically it can be believed with some degree of confidence. With the Average Nearest Neighbor tool, an index is calculated that reflects the average distance from a feature to all its neighbors compared to the average distance for a random distribution.

Learning objectives

- *Test for statistical significance*
- *Evaluate z-scores*
- *Understand random distribution*
- *Understand clustering versus dispersion*

Preparation

- *Read pages 63–79 and 88–96 in The ESRI Guide to GIS Analysis, Volume 2.*

Introduction to spatial statistics

Chapters 8 and 9 deal with spatial statistics. No longer are the commands a simple step to making an output like a buffer or a density. This material gets into a higher level of mathematics and begins to delve into the theoretical realm. With spatial statistics, you deal with the mathematical analysis of existing data to predict the possibility that something will or will not happen. The formulas are complex, and understanding the steps can be difficult. But the results are a less subjective way to confirm what the data is representing.

The ESRI Guide to GIS Analysis, Volume 2, is the basis for these chapters, and should be referenced for a more complete explanation of the formulas used in the spatial statistics tools. The descriptions and charts it contains provide a good background for the tools and explain how the tools fit into the realm of statistical analysis.

Tutorials 8 and 9 deal with many of the spatial statistics tools available at all license levels of ArcGIS Desktop software. Understanding when to use which tool, and why, can be confusing, so a general description of each tool's function is given here.

The first four tools are used to answer a global question: "What is the probability that the distribution of these features is occurring due to random chance?" These tools don't create anything that can be symbolized on your map, although it may be helpful to capture some of the graphics that they create. The tools are important, however, in establishing a statistical foundation and a statistical confidence level for your analysis.

Average nearest neighbor (clustering by location). The average nearest neighbor index looks at each feature and the single nearest feature, then calculates a mathematical index. It then creates a hypothetical, randomly distributed set of data and calculates the index again. The degree of clustering is measured by how much the index for the real data differs from the index of the hypothetical data. Use this tool to see if the physical locations are closer together than would be expected with a random distribution.

Getis-Ord General G (clustering by value). The General G statistic looks at the similarity of the values associated with the features within a critical distance of each other. Areas where the values are similar have a strong clustering, whether they be high values or low values. Use this tool to see if there are areas where similar values are closer together than would be expected with a random distribution.

Multidistance clustering, or Ripley's K function (clustering by location but using multiple features and multiple distances). Similarly to the nearest neighbor method, the K Function looks at the distance from a feature to a large number of the nearest features to determine a clustering index. It may also be run for multiple distances to see which distance produces the most significant clustering. This method can detect a more subtle clustering effect than can be detected with the nearest neighbor method. Use this tool to determine if physical locations are clustering due to factors beyond the next nearest feature.

Spatial autocorrelation, or global Moran's I (clustering by both location and value). Spatial autocorrelation determines if there is an underlying geographic clustering of the data based on both location and attribute value. Use this tool when the physical location data has an attribute associated with it that may be influencing the clustering.

The next two tools can help pinpoint the locations of clustering patterns.

Cluster/outlier analysis, or Anselin local Moran's I. Local Moran's I method identifies areas of clustering by location as well as by values similar in magnitude. Use this tool when you desire a graphic output on the map, and the suspected clustering is due to both location and an attribute associated with the features.

Getis-Ord hot spot analysis, or Gi* (clustering of high and low values). Gi* indicates areas where values associated with the features are clustering. A positive index shows clusters of high values (hot spots); a negative index shows clusters of low values (cold spots). Use this tool to create a map display of the locations of hot spot and cold spot clusters based on values.

8-1
8-2
8-3
8-4

Within this framework are some general terms and values that are derived from the world of statistical analysis. Terms like z-score, null hypothesis, confidence level, and statistical significance are described in the *The ESRI Guide to Spatial Analysis, Volume 2*, and in ArcGIS Desktop Help under Geoprocessing.

It is also highly recommended that you view the free ESRI Virtual Campus presentation on spatial statistics. Go to the training gateway for more information: `http://training.esri.com`.

Introduction

The Average Nearest Neighbor tool is used to determine if a set of features shows a statistically significant level of clustering or dispersion. It does this by measuring the distance from each feature to its single nearest neighbor and calculating the average distance of all the measurements. The tool then creates a hypothetical set of data with the same number of features, but placed randomly within the study area. Then an average distance is calculated for these features and compared to the real data. A nearest neighbor index is returned, which expresses the ratio of the observed distance divided by the distance from the hypothetical data. If the number is less than one, then the data is considered to exhibit clustering; if it is more than one, then the data is exhibiting a trend toward dispersion.

The idea behind this tool is that things near each other are more alike than things occurring far apart, commonly called Tobler's First Law of Geography. If the distances between some features are significantly smaller than the distances between other features, the features are considered to be clustered.

The calculation performed by this tool is best done against point data. It can be run against line and polygon data, but the centroids of the features will be used. Not only will this not represent the data very well, but the hypothetical distribution will also be flawed when it is compared with the real data.

The other thing to be careful of is the area value the tool will use for the calculations. The default is to use the area of a rectangle defined by the data's outlying values. As an alternative, a nonrectangular measured area can be used.

It is most appropriate to calculate the nearest neighbor index for equal study areas. The method works best for comparing data occurring in the same area over time rather than data occurring in different areas.

In addition to the nearest neighbor index, the tool will calculate a z-score. This number represents a measure of standard deviation that can help you decide whether or not to reject the null hypothesis.

For pattern-analysis tools, the null hypothesis states that there is no pattern. When you perform a feature pattern analysis and it yields either a very high or a very low z-score, this indicates it is very unlikely that the observed pattern is the product of a random distribution.

In order to reject or accept the null hypothesis, you must make a subjective judgment regarding the degree of risk you are willing to accept for being wrong. This degree of risk is often given in terms of critical values or confidence level. At a 95 percent confidence level the critical z-score values are -1.96 and +1.96 standard deviations. If your z-score is between -1.96 and +1.96, you cannot reject your null hypothesis; the pattern exhibited is a pattern that could very likely be the result of a random distribution. However, if the z-score falls outside that range (for example -2.5 or +5.4), the pattern exhibited is probably too unusual to be random. A statistically significant z-score makes it possible to reject the null hypothesis and proceed with figuring out what might be causing the clustering or dispersion.

Scenario   The Fort Worth Fire Department has again asked for your help with a spatial analysis project. It would like to know if the EMS calls for Battalion 2 have a tendency to cluster. If so, it may consider stationing the Emergency Intensive Care Units (EICU) at locations near the hot spots. You will need to use the Average Nearest Neighbor tool to analyze the data and determine the degree of clustering or dispersion. Ultimately, you will need to decide whether or not to reject the following null hypothesis:

> EMS calls for service in February 2007 are randomly distributed across the study area.

Data   The EMS Calls–Feb07 data is the same response data from the previous tutorials, with a definition query built to extract only the calls with an incident type GRP value between 30 and 39. This group represents the EMS calls only.

Also included is a layer with the Battalion 2 boundary. This is a new layer with only one polygon. It will be used to determine the area of study and to make sure that the other battalions are not accidentally included in the study. Remember that the Average Nearest Neighbor tool is very sensitive to the size of the study area.

Examine data

You'll start with the EMS calls for Battalion 2. By doing quick visual analysis as you did earlier in the book, you can see some groupings of features. You will then analyze the data statistically to assess the extent of clustering. This will add a level of credibility to your reports that purely visual analysis sometimes lacks.

8-1

8-2

8-3

8-4

1 In ArcMap, open Tutorial 8-1.mxd. The Average Nearest Neighbor tool is very sensitive to the area of study, so to ensure the best results you will want to look at and record the measured area of the Battalion2 layer.

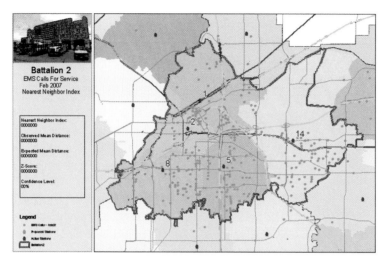

2 Open the attribute table for the Battalion2 layer and make a note of the square footage in the Shape_Area field.

3 Open the properties of the EMS Calls–Feb 07 layer and examine the definition query that has been created.

With this information in hand, you are ready to start pattern analysis. This analysis will return a nearest neighbor index and a z-score on the basis of which you will decide if the data clusters.

4 Close the Layer Properties dialog box.

Analyze patterns with the Average Nearest Neighbor tool

1 Use the Search window to locate and open the Average Nearest Neighbor tool.

2 Enter parameters as follows:

- **Input Feature Class: EMS Calls–Feb 07**
- **Area: 520175356 (square feet)**
- **Check the Generate Report box**

3 When your dialog box matches the image, click OK.

4 When the process is complete, click the pop-up box to display the results. If you miss the pop-up box, you can open the results window from the main menu using Geoprocessing > Results. Make a note of the following values from the results:

- **Observed mean distance** _____
- **Expected mean distance** _____
- **Nearest neighbor ratio** _____
- **Z-score** _____

The tool will calculate the distance from each feature to its nearest neighbor. Then the average of all these distances is found. Next, the tool creates a hypothetical random distribution of points using the same number of features and the same study area and repeats the calculations. The option to provide an area ensures that the hypothetical random dataset covers the same area as your study area.

5 In the results dialog box, double-click the HTML Report File to open the graphic display of the results. The graphic depicts a bell curve showing the amount of clustering the data is displaying. It also reports the nearest neighbor ratio (index) and the z-score.

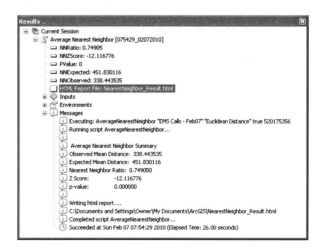

8-1
8-2
8-3
8-4

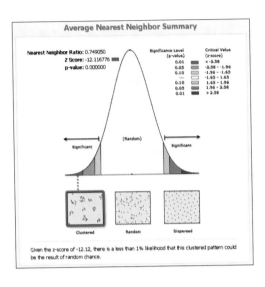

A set of features that occurs in a totally random manner would have an index of 1. A set of features showing more dispersion than the hypothetical data would have an index greater than 1. A set of features showing more clustering than the hypothetical data would have an index less than 1.

Your dataset returned an index of 0.75, which tells us that the features are trending toward clustering.

The z-score is –12.12 with a significance level of 0.01. This score gives us the confidence level of rejecting the null hypothesis. Scores closer to 0 will have a low confidence that the distribution is not random. The greater values, either in the positive or negative direction, give a higher confidence that the distribution is exhibiting clustering or dispersion.

The z-score of –12.12 tells us that there is a 99 percent confidence that the data distribution is not due to random chance (this is the reverse of the statement in the dialog box that there is less than a 1 percent chance that this clustered pattern could be the result of random chance). With this high confidence level, you can safely reject the null hypothesis.

6 Close the Average Nearest Neighbor graphic display and the results window.

7 The Average Nearest Neighbor tool doesn't create any additional features; it only reports back the values from the requested calculations. To record these on the map, edit the text box on the title block and add the results.

As you work through the other tutorials, try making a box plot to include in your layout.

8 Save your map document as **Tutorial 8-1.mxd** in the \GIST2\MyExercises folder. If you are not continuing on to the exercise, exit ArcMap.

Box plot

Statisticians will often look at a histogram of the data to see the distribution, as you did in an earlier tutorial. Another standard display of data distribution is the box plot. Box plots can be made using the View > Graphs > Create tool and setting the graph type to Box Plot. Select a layer and the value field, and the graph will be created. The image to the right shows an example of a box plot of the data from this tutorial with the FEE field as the value, and a diagram of how to read a box plot is shown in the image below.

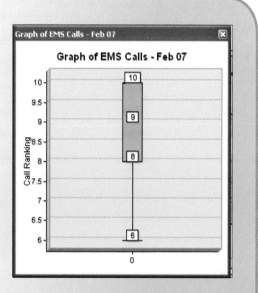

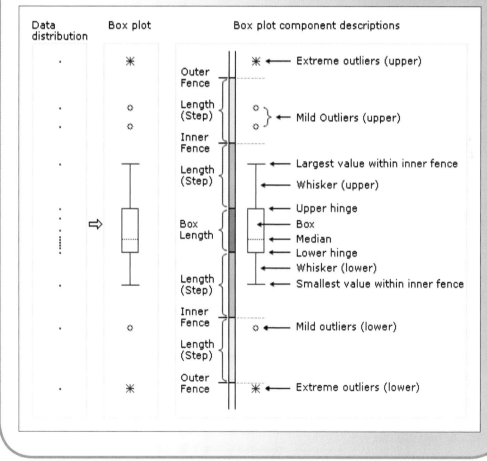

Exercise 8-1

The tutorial showed how to calculate the nearest neighbor index and a z-score. It also explained how to interpret the results.

The fire department now wants to see if false alarms are exhibiting any clustering. If so, the department will target its safety campaign to those areas, explaining how to recognize a real emergency and how to avoid false alarms.

- Continue using the map document you created in this tutorial or open Tutorial 8-1.mxd from the \GIST2\Maps folder.
- Turn off the EMS Calls Feb–07 layer.
- Turn on the Incident_Feb07 layer.
- Create a definition query to restrict the data where the field inci_type is between 700 and 745 inclusively: [inci_type] >= '700' AND [inci_type] <= '745'. Also included are the incident types 7351 and 7352.
- Perform the Average Nearest Neighbor calculations.
- Record the results in the text box on the map.
- Save the results as **Exercise 8-1.mxd** in the \GIST2\MyExercises folder.

WHAT TO TURN IN

If you are working in a classroom setting with an instructor, you may be required to submit the maps you created in tutorial 8-1.

Turn in a printed map or screen capture of the following:

Tutorial 8-1.mxd

Exercise 8-1.mxd

Tutorial 8-1 review

The Average Nearest Neighbor tool was used to analyze the fire department data to test for clustering. It returned a set of numbers that gave a confidence level of the statistical analysis. That value, called the z-score, gave a probability of 99 percent that the clustering was statistically significant.

The tool is sensitive to area. Remember that the nearest neighbor index cannot be compared between areas of different sizes, so it is most valuable to study results in the same area over time.

Once a set of features is determined to be clustered at a significant level, and the null hypothesis is rejected, you can move on to hot spot or clustering analysis tools.

The Average Nearest Neighbor tool doesn't use additional weight fields. The index is only affected by the next nearest feature within the study area.

STUDY QUESTIONS

1. Why does an irregular area create problems for the Average Nearest Neighbor tool?

2. Would running the Average Nearest Neighbor tool several times produce the same numbers? Why or why not?

Other real-world examples

A pizza chain might determine the clustering of their clients using the nearest neighbor index. This would let the company identify customers who are close to each other, more so than would occur at random, and perhaps categorize the cluster as a delivery hot spot.

An epidemiologist might look for clusters of a certain illness and match the clusters to the locations of food sources or water to determine if these other geographic elements are associated with an underlying cause.

8-1
8-2
8-3
8-4

Tutorial 8-2

Identifying the clustering of values

With some features, the locations alone are not the factor that determines their clustering. It is instead the values associated with the features. The Getis-Ord General G tool evaluates a value associated with the features in the data and determines whether the high values or the low values cluster.

Learning objectives

- *Test for statistical significance*
- *Evaluate z-scores*
- *Understand random distribution*
- *Understand clustering of values*

Preparation

- *Read pages 104–108 in* The ESRI Guide to GIS Analysis, Volume 2.

Introduction

The General G statistic measures concentrations of high or low values over the entire study area. It measures within a specific distance how high or low the values are, and compares this to how high or low the rest of the values are. It is termed General because it deals with the entire study area rather than a localized area.

This tool assumes that features have an association because they have similar attribute values, as well as being within the specified distance of each other. The input features for this tool should have an attribute demonstrating some characteristic or value associated with the location. The result will determine if the values are clustered, not the locations. Additionally, it will distinguish between clustering of high values versus low values.

Another aspect of this tool is the distance over which the General G statistic will be determined to be significant. The "conceptualization of spatial relationships" setting specifies the type of relationships the features have to one another. To set this correctly, you must understand the internal relationships of the data.

If one feature's value has an effect on the values of the nearby features (as with measurements of rain, for example), then the inverse distance setting can be used. To amplify this relationship, use the inverse distance squared setting.

For features such as calls for service, it is best to use the fixed distance band setting and run the tool several times to determine how location is affecting the clustering. This is because one call for service has no effect on the probability that another call for service will occur in the same area. In other words, a call for service to fight a kitchen fire today will have no relationship to a call for an auto accident on the next block that may happen next week.

Some data that does have a spatial effect from one feature to another might benefit from a combination of the two settings. The zone of indifference setting states that features within a specified distance have an impact on the other features, and after that the influence quickly drops off. To get the best results, either use the fixed distance band setting, or if you can determine that there is a geographic influence among the features, use the zone of indifference.

The General G tool can be run multiple times for different distances. The highest z-score corresponds to the point of maximum clustering.

Scenario The fire department would like to know if the calls ranked as a high priority are clustering. Two things that the previous analysis didn't show were the effect of the FEE ranking on the clustering and the distance at which the clustering occurs. This distance is critical in seeing the compactness of the groupings, and will be used later in the hot-spot analysis.

You will run the analysis at various distances and determine the maximum z-score, confidence level, and distance band for clustering.

This is the null hypothesis for this analysis:

The Priority Ranking values for the features in Calls for Service–Feb 07 are randomly distributed across the study area.

On the basis of your analysis, you will decide whether or not to reject the null hypothesis.

Data The Calls for Service–Feb07 data is the same response data from the previous tutorials. The field FEE contains ranks from 1 to 10, reflecting each call's priority.

Additional layers are included for background interest.

8-1
8-2
8-3
8-4

Investigate threshold distances

1 **In ArcMap, open Tutorial 8-2.mxd.** Once again, here's a map of the Battalion 2 calls for service. Even though you might see some groupings, you can't see what effect the call priority has. Visual analysis can't provide that, so you instead need to rely on spatial statistics for the answer.

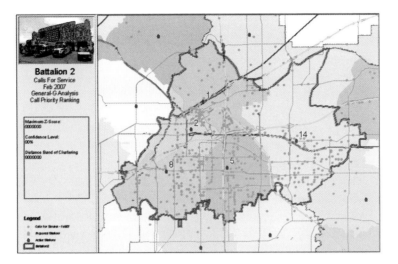

You will be running the General G tool several times to find the value at which the z-score peaks, but what values should you try? If you don't get within the right range, you may not discover the clustering. To find the distance range to try, you'll use a combination of visual analysis and a statistics utility.

2 Zoom in to an area that exhibits some visual grouping. A quick scan shows about seven or eight calls located in this area.

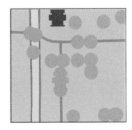

You will use the Calculate Distance Band from Neighbor Count along with the observed number of neighbors from your visual inspection to get an idea of what distance to enter into the General G tool and start recording z-scores.

3 Use the Search window to locate and open the Calculate Distance Band from the Neighbor Count tool.

4 Use Calls For Service–Feb 07 as the input features and set the number of neighbors to **7.** Click OK. The tool will find the minimum, average, and maximum distances at which each feature can find at least seven neighbors.

5 Open the Results box when the process is complete and note the minimum, average, and maximum distance. You will base your search distances on the average. Close the results box.

The average distance to find seven neighbors was around 1,000 feet, which you should bracket high and low to determine which values to use with the General G tool. The distances to try in the General G tool should be from 200 feet to 1,200 feet at 200-foot intervals. Since the map units are feet, these numbers can be entered without adding units, but if the map units were set to something else, you could enter the abbreviation for feet in the input box. Note the z-score at each of the distances and watch for the peak value in this range. The distance that produces the largest z-score will be the distance with the most significant clustering.

8-1
8-2
8-3
8-4

Run the High/Low Clustering (Getis-Ord General G) tool

1 Use the Search window to locate and open the High/Low Clustering tool.

2 Enter the parameters as follows:

- **Input Feature Class: Calls for Service–Feb 07**
- **Input Field: FEE**
- **Check the Generate Report box**
- **Conceptualization of Spatial Relationships: Fixed Distance Band**
- **Distance Band: 200**

3 When your dialog box matches the image, click OK.

4 Double-click the HTML Report File in the results window to view the graphic report. The result shows that the data does have a degree of clustering at this distance. But is this the distance at which the greatest clustering is found? Also note that many features did not have neighbors at this distance, which makes them invalid in the calculation. A better distance would have much fewer features with no neighbors so that the maximum number of features are included in the calculation.

5 Record the General G index and the z-score for 200 feet. Close the results dialog boxes.

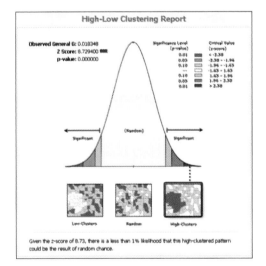

YOUR TURN

Run the High/Low Clustering tool five more times, increasing the distance by 200 feet each time. Record the G indices and the z-scores for distances of 400, 600, 800, 1,000, and 1,200 feet. You might want to list your results in a table like this:

Distance	G index	Z-score
200		
400		
600		
800		
1,000		
1,200		

Graphing the results will show where the z-score peaks, which is the point of significant clustering of high values. Remember that this is showing the distance at which values are clustering at a rate higher than expected by chance. Notice also that as the distances increase, the number of features with no neighbors decreases.

6 Update the text box in the title bar with the maximum z-score, confidence level, and distance band of clustering. Print or make a screen capture of the map.

7 Save your map document as **Tutorial 8-2.mxd** in the \GIST2\MyExercises folder. If you are not continuing on to the exercise, exit ArcMap.

8-1
8-2
8-3
8-4

Exercise 8-2

The tutorial showed how to calculate a z-score for assessing whether the values of features are clustered. The fire department would like you to do the same clustering analysis for another set of data.

- Continue using the map document you created in this tutorial or open Tutorial 8-2E.mxd from the \GIST2\Maps folder.
- Turn off the Calls for Service–Feb07 layer and turn on the Calls For Service–Jan 07 layer.
- Perform the high/low clustering calculations for the Calls for Service–Jan 07 layer. Use the same distance parameters as in the tutorial.
- Record the z-scores for each distance and determine which one provides the highest z-score.
- Record the results in the text box on the map border.
- Change the titles, colors, and legend to make a visually pleasing map.
- Save the results as **Exercise 8-2.mxd** in the \GIST2\MyExercises folder.

WHAT TO TURN IN

If you are working in a classroom setting with an instructor, you may be required to submit the items you created in tutorial 8-2.

Turn in a printed map or screen capture (together with your z-score tables) for the following:

Tutorial 8-2.mxd

Exercise 8-2.mxd

Tutorial 8-2 review

The values for the features were used to demonstrate clustering. By running the tool several times with a range of distances, the highest z-score was found. This peak z-score corresponds to the distance at which the clustering of values is the strongest.

STUDY QUESTIONS

1. What other types of values might be used for this analysis?
2. How did this analysis differ from the average nearest neighbor analysis?

Other real-world examples

A retail chain might determine the clustering of customers who have spent a large amount of money in the stores to set up a home-based shopping experience.

A map of crop yield might be used to determine clustering of higher-than-expected harvesting numbers. This may in turn be used to study factors contributing to high growth.

Oceanographers may look for clustering of measured salinity levels in the ocean and match the results to currents to see if there is any correlation.

8-1
8-2
8-3
8-4

Tutorial 8-3

Checking for multidistance clustering

Multidistance spatial cluster analysis, also known as Ripley's K function, examines the counts of neighboring features at several distances. If the count is higher than what would occur in a random distribution, then the features are considered clustered. A graph is automatically created for the tested distances.

Learning objectives

- *Test for statistical significance*
- *Evaluate z-scores*
- *Understand random distribution*
- *Evaluate clustering of values at multiple distances*

Preparation

- *Read pages 97–103 in* The ESRI Guide to GIS Analysis, Volume 2.

Introduction

The Ripley's K function method of pattern analysis measures the distance between features to determine clustering. But unlike the average nearest neighbor index, it includes all the neighboring features in the calculation, not just the nearest one.

As with other methods, the K function generates a hypothetical random distribution using the same number of features and the same area. The calculation is done on both the real data and the hypothetical data, and the difference between the observed index value and the index value generated by the hypothetical random data indicates the degree of clustering. When plotted on a graph, the area with the greatest distance between the two values is the point of most significant clustering.

The index is calculated by measuring the distance from each feature to all the other features in the dataset; then a mathematical operation generates a K value. This is repeated for various user specified distances and plotted on a graph. Then a hypothetical set of data is created by randomly tossing the same number of features into the same study area and performing the calculation again. A starting distance and an increment distance are entered for the tool to run through a range of distances, eliminating the need to run the tool multiple times and manually plot the results.

Using the optional permutations, the calculation is performed against the random dataset multiple times, creating a confidence envelope. This will represent on the graph the highest and lowest values achieved in the random datasets created in the multiple permutations. The peak at where the observed K index exceeds the confidence envelope the most is the point at which the greatest clustering occurs.

Because the distances between all features are used in the K function calculation, the geographic shape of the area has a great impact on the results. The hypothetical random dataset is created in the same area and must be subject to the same shape and distance as the real data. It is recommended that a single polygon representing the study area be used in the calculations.

Another feature of multidistance clustering is the option to include a weight. As with the other tools, the weight value will give some features more importance in the calculations.

Scenario The response call dataset was proven to have significant clustering with the pattern analysis tools you have used. The Multi-Distance Spatial Cluster Analysis tool will take into account the specific study area, and include the relationships of each feature to all the other features. This, along with the weighting factor, will produce a better result.

You will run the analysis for a set of distances and determine the maximum z-score, confidence level, and distance band for clustering.

This is the null hypothesis for this analysis:

> The features in the Calls for Service–Feb 07 dataset, when weighted with priority rankings, are randomly distributed across the study area.

On the basis of your analysis, you will decide whether or not to reject the null hypothesis.

Data The Calls for Service–Feb07 data is the same response data from the previous tutorials. A field containing FEE values contains ranks from 1 to 10, reflecting each call's priority.

Additional layers are included for background interest.

8-1
8-2
8-3
8-4

Run the Multi-Distance Spatial Cluster Analysis (Ripley's K Function) tool

1 In ArcMap, open Tutorial 8-3.mxd. Once again, here's a map of the Battalion 2 calls for service. Ripley's K function will be used to analyze the patterns for this dataset, in relation to the unique shape of Battalion 2. The features that fall outside Battalion 2 will have an effect on the data both by the distance they are from the other features and their weight values.

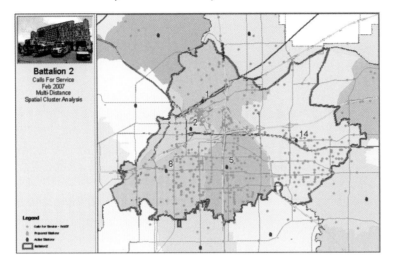

2 Use the Search window to locate and open the Multi-Distance Spatial Cluster Analysis tool. There are many parameters and options to set with this tool, and each can have a large impact on the results. The ones used here are based on the distances used for the high/low clustering calculations and will encompass the value that showed the best clustering.

3 Set Input Feature Class to Calls for Service–Feb 07 and set the location for the output table to \GIST2\MyExercises\MyData\KFunctionFeb07.

The output table will store all the various outcomes of the calculations. After the tool completes the calculations, this data can be used to make a graph that can be included on the map.

4 Fill out the next portion of the dialog box as follows:

- **Number of Distance Bands: 10**
- **Check the Display Results Graphically box**
- **Weight Field: FEE**
- **Beginning Distance: 200**
- **Distance Increment: 100**

The Number of Distance Bands determines how many measurement rings to make. The Distance and Increment values give a starting value, and how large each ring will be, respectively. With the values you entered, the first ring will be at 200 feet, then nine more rings 100 feet apart will be used. As with other analyses, the calculation will be weighted based on FEE values, giving call priority ranking an influence on the results. The optional Confidence Envelope is left at zero for now, but you will use it later in the tutorial.

5 Set Boundary Correction Method to None, and Study Area Method to Minimum Enclosing Rectangle. (If you are using an ArcInfo license, set the Study Area Method to "User provided Study Area Feature Class," and Study Area Feature Class to Battalion2. This will give better results, but will be different from the results shown here.)

The Boundary Correction Method can be used to help simulate features falling outside the study area. This can give the features nearest the edge simulated neighbors to prevent a black hole effect. For this dataset, there are features falling outside the study area, so no boundary correction is necessary. If the dataset had been cut at the boundary from a larger dataset, then a boundary correction would be appropriate.

As mentioned before, this tool is very sensitive to the study area being provided, especially if multiple permutations are used. The polygon representing Battalion 2 was selected and exported to a separate feature class to use with this calculation. Setting the study area method to "User provided" allows this feature class to be used for all the calculations. If you have an ArcInfo license, be sure to use this option.

6 Click OK. When the process finishes, open the results window. The results are displayed in the Results window, and all the values are stored in a table called KFunctionFeb07, which is added to your table of contents.

7 Close the Results window.

Create a graph of the Difference field

The distance of the most significant clustering will be where the observed values exceed the expected values by the greatest margin. You can show this in a graph.

1 At the top of the table of contents, click List by Source. Right-click KFunctionFeb07 and select Open to display the table. Click Table Options and select Create Graph.

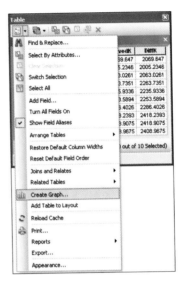

2 In the Create Graph wizard, set Graph type to Vertical Line, set Y field to DiffK, and set X label field to ExpectedK. When your dialog box matches the image, click Next, then click Finish to create the graph. Close the attribute table.

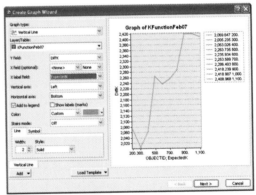

3 Right-click the graph and select Add to Layout. Close the graph and the attribute table.

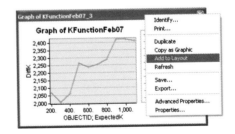

4 Resize the graph in the layout and move it to the title bar. The graph shows a peak in the values at 900 feet, but there was another peak at 500. Any of these might be the point of the most pronounced clustering.

This set of calculations didn't include the confidence envelope that the permutations option would provide. The confidence envelope is developed by running hypothetical random distributions many more times, and plotting the highest and lowest values for each distance. The values of the Difference field can then be compared to a larger sampling of possible random data. The value that exceeds the confidence envelope by the greatest margin will be the distance of most significant clustering.

Legend

- Calls For Service - Feb07
- Proposed Stations
- Active Stations
- Battalion2

Rerun the Multi-Distance Spatial Cluster Analysis tool with a confidence envelope

1 Open the Multi-Distance Spatial Cluster Analysis tool.

2 Set the following parameters:

- **Input Feature Class: Calls For Service–Feb07**
- **Output Table: \MyExercises\MyData.mdb\KFunctionFeb07ConfEnv**
- **Number of Distance Bands: 10**
- **Compute Confidence Envelope: 99 Permutations Note:** This many permutations may take 15 to 20 minutes to complete. For faster results, you may try using only 9 permutations. This will still calculate a confidence envelope, although your results will look slightly different from what you see here.
- **Check the Display Results Graphically box**
- **Weight Field: FEE**
- **Beginning Distance: 200**
- **Distance Increment: 100**
- **Boundary Correction Method: None**
- **Study Area Method: Minimum Enclosing Rectangle**

8-1
8-2
8-3
8-4

3 When your dialog box matches the image, click OK. The tool will calculate the K index for the real data, then calculate the index for random hypothetical data 99 times. With this expanded set of data with which to compare the real value, it will be easier to see which clusters are unlikely to occur randomly.

4 Note the results of the K function, then close the Results window. A table containing the results will be added to the table of contents (on the List by Source pane).

Examine results

To find the distance of maximum clustering, it is best to calculate the difference between the observed K index and the upper limit of the confidence envelope.

1 Open the KFunctionFeb07ConfEnv table. Create a new floating point field named **Difference** in the table and calculate the following value: **[ObservedK]– [HiConfEnv]**. Click OK.

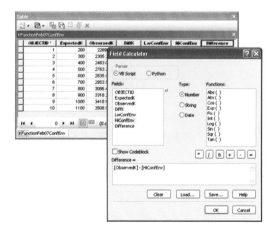

2 Make a graph as previously shown using Difference as the Y field and ExpectedK as the X label field. (Your results may differ depending on the number of permutations used.)

3 Add the graph to the layout, then close the graph and attribute table.

4 **Resize the graph and add it to the title bar.** The resulting graph makes it pretty clear that the 500 value is the peak of the graph. The difference between the two was the use of the confidence envelope and the 99 permutations to check a larger sample of the hypothetical random data.

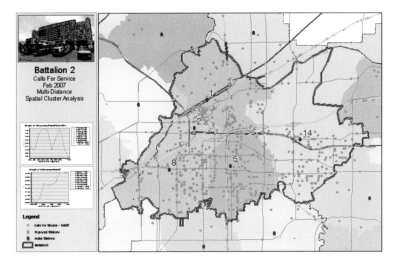

5 Save your map document as **Tutorial 8-3.mxd** in the \GIST2\MyExercises folder. If you are not continuing on to the exercise, exit ArcMap.

Exercise 8-3

The tutorial showed the use of the Multi-Distance Spatial Cluster Analysis tool for calculating a K function index and a confidence envelope in order to assess the significance of clustering.

The fire department would like you to do the same analysis for another set of data.

- Continue using the map document you created in this tutorial or open Tutorial 8-3E.mxd from the \GIST2\Maps folder.
- Perform the multidistance cluster analysis calculations using a confidence envelope for the Calls for Service–Jan07 layer. Use the same distance parameters as in the tutorial.
- Create a graph from the data and add it to the map.
- Change the titles, colors, and legend to make a visually pleasing map.
- Save the results as **Exercise 8-3.mxd** in the \GIST2\MyExercises folder.

WHAT TO TURN IN

If you are working in a classroom setting with an instructor, you may be required to submit the items you created in tutorial 8-3.

Turn in a printed map or screen capture of the following:

Tutorial 8-3.mxd

Exercise 8-3.mxd

Tutorial 8-3 review

The spatial statistics tools from the previous tutorials use a standard formula for computing the z-score, and they can be run many times to test data. With Ripley's K function, permutations are used to run a large number of scenarios, basing the results on many more tests of random hypothetical data.

Because so many random samples are used, the area in which they are generated is very critical for making the hypothetical data as similar to the real data as possible. If the random data is distributed in a much larger space, they will not simulate reality very well, thus giving misleading results.

Finding the distances to use for the tool can be rather subjective. If you have knowledge of the data, and know some of the distances that may affect the data, then these might be the basis for the analysis. For instance, if you are looking for clusters of commute distances and you know that the average commute is fifteen miles, then this may guide you in selecting distances.

The data can cluster at different distances. The confidence envelope provides a basis for comparing the levels of clustering and determining the most significant among them.

STUDY QUESTIONS

1. How is the confidence envelope calculated?
2. What does the confidence envelope tell you?
3. Would you expect to see a difference between 9 and 999 permutations? Why?

Other real-world examples

Ripley's K function may be used to examine the clustering of a certain plant species. A regional search may be done for the significance of clustering rather than only the next nearest neighbor.

A catalog sales company may look for clustering of customers, but on a more regional level. This clustering will take into account other customers across a broader distance and reveal more subtle patterns.

8-1
8-2
8-3
8-4

Tutorial 8-4

Measuring spatial autocorrelation

Measuring spatial autocorrelation involves not only using feature locations or attribute values alone, but using both simultaneously. Given a set of features, a Moran's I index is calculated to determine if the data is clustered, random, or dispersed. Moran's I requires a weight either from an attribute or as a count of features in a defined area.

Learning objectives

- *Test for statistical significance*
- *Evaluate z-scores*
- *Understand random distribution*
- *Evaluate clustering of values and distances*

Preparation

- *Read pages 80–87 and pages 118–126 in* The ESRI Guide to GIS Analysis, Volume 2.

Introduction

The Moran's I function is designed to find clustering using the attribute values as well as locations of features. This is typically done with polygons that contain a summary statistic, such as census data or density data. It's important to note that the Moran's I function doesn't identify clusters on the map, but rather identifies whether the pattern of values across the study area tends to be clustered, random, or dispersed.

The point data of calls for service will need to be aggregated into polygons for this analysis. This will give us clusters of density. One way to create summary polygon data is to overlay a grid of a set size and count the number of locations that fall within each polygon. ArcMap contains a command to create the grid; then a spatial join can be performed to count the features in each grid cell.

Determining grid cell size is critical to the analysis. Since the tool is looking for clustering of densities, it is important to choose a grid size that will have polygons with at least one point in them, but also provide a larger range of values across the dataset. If you have some idea of the dataset's distance characteristics, this can be used to set grid size. For the calls for service data, using the size of a standard city block, or 500 feet, would be acceptable.

Once the grid is created and the spatial join is performed, the Moran's I tool will be run using the output grid as the features and the count of call locations as the attribute value.

The Moran's I tool compares the values for neighboring features. A comparison is made of the differences in values between each pair of neighbors and all the other features in the study area. If the average difference between neighboring features is less than between all the features, the values are considered clustered.

As with the other spatial statistics tools, the most significant clustering will occur at the point where the z-score peaks. The command can be run several times to find the peak z-score.

Scenario   The fire department is looking at density of calls per block and would like to see at what distance these densities cluster.

The first step will be to overlay a grid onto the point data and count the number of calls in each cell. Then the Moran's I tool can be run with a range of distances to determine at what point the features cluster.

This is the null hypothesis for this analysis:

Calls for service are randomly distributed among city blocks.

Data   The Calls for Service–Feb 07 data is the same response data from the previous tutorials. Additional layers are included for background interest.

Aggregate data

1   **In ArcMap, open Tutorial 8-4.mxd.** Once again, here's a map of the Battalion 2 calls for service. A grid will be overlaid and used to count the features, creating a density grid. Then the Spatial Autocorrelation Moran's I tool can be used to look for clustering of the density values.

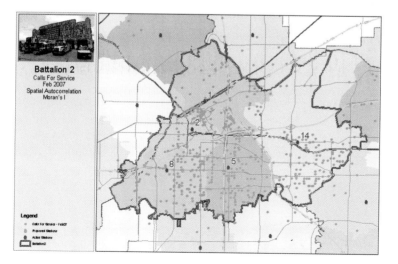

8-1
8-2
8-3
8-4

2 Turn on the layer 500 ft Grid. The layer is a 500-by-500 foot grid. Creating a spatial join between the grid layer and the Calls for Service layer will output a layer with the same grid and a count of how many call locations occur inside each grid cell.

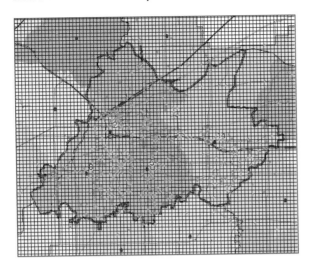

3 Right-click the 500 ft Grid layer and select Joins and Relates > Join. Complete the Join Data dialog box as follows:

- Under "What do you want to join to this layer?", select "Join data from another layer based on spatial location"
- For item 1, set the join layer to Calls for Service–Feb07
- For item 3, set the output file to \MyExercises\MyData.mdb\ Calls_for_Service_Feb07_Grid

4 When your dialog box matches the image, click OK. The polygon file that was created now shows the density of the calls for service. Before running the Spatial Autocorrelation tool, you should remove the cells that have no value. This can be accomplished with a definition query.

5 Turn off the 500 ft Grid layer.

6 Open the properties of the Calls_For_Service_Feb07_Grid layer and set a Definition Query to allow only Count values that exceed zero. Click OK and OK again.

The result displays only the grid cells that contain at least one call for service. While you may look at this and think you see clustering, you cannot see the values associated with the grid cells that will be used in the spatial autocorrelation calculation.

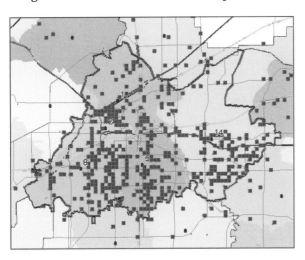

Run the Spatial Autocorrelation tool

The Spatial Autocorrelation tool will use familiar settings from the other spatial statistics tools. You will enter input data to analyze. Then you will select a value field and a Conceptualization of Spatial Relationships method. The tool will be run for a range of values to determine the peak z-score. With the grid set at 500 feet, the search for the peak z-score can begin at 1,000 feet and increase by 100 feet for each round.

1 Locate and open the Spatial Autocorrelation tool.

2 Enter the following parameters in the dialog box:

- Input Feature Class: Calls_For_Service_Feb07_Grid
- Input Field: Count_
- Check the Generate Report box
- Conceptualization of Spatial Relationships: Zone of Indifference
- Distance Band: 1000 (feet)

8-1
8-2
8-3
8-4

3 When your dialog box matches the image, click OK.

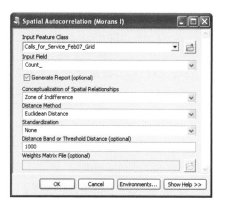

4 Record the resulting z-score from the results window.

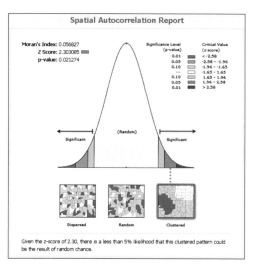

YOUR TURN

Repeat the process for the distances shown below. Keep all the other parameters the same, changing only the Distance Band. You might want to list your results in a table like this:

Examine the z-scores and find the highest. Check the confidence level to determine if the block level data is clustered. Based on the results, can you reject the null hypothesis?

Distance	Z-score
1,000...............	
1,100...............	
1,200...............	
1,300...............	
1,400...............	
1,500...............	
1,600...............	
1,700...............	
1,800...............	
1,900...............	

5 Save your map document as **Tutorial 8-4.mxd** in the \GIST2\MyExercises folder. If you are not continuing on to the exercise, exit ArcMap.

Exercise 8-4

The tutorial showed the use of the Spatial Autocorrelation tool to determine Moran's I values and z-scores for assessing the level of the clustering.

The results may be different if a different grid size is used. Repeat the analysis using a 200-foot grid and determine if you can reject the null hypothesis.

- Open Tutorial 8-4E.mxd.
- Perform a spatial join of the 200-foot grid with the Calls for Service–Feb07 data.
- Create a definition query as you did in the tutorial.
- Run the Spatial Autocorrelation tool for the new grid layer using distances from 250 to 600 feet in increments of 50 feet. Use the same parameters as in the tutorial.
- Record the z-scores with their confidence levels and decide if you can reject the null hypothesis.
- Change the titles, colors, and legend to make a visually pleasing map.
- Save the results as **Exercise 8-4.mxd** in the \GIST2\MyExercises folder.

WHAT TO TURN IN

If you are working in a classroom setting with an instructor, you may be required to submit the items you created in tutorial 8-4.

Turn in a printed map or screen capture, with z-scores, for the following:

Tutorial 8-4.mxd

Exercise 8-4.mxd

8-1
8-2
8-3
8-4

Tutorial 8-4 review

Moran's I measures the clustering of value densities. The locations of features and their values are used in the calculations, and as with the other pattern analysis tools the z-score is tracked as a measure of confidence in the clustering. The fire department calls for service data used in the tutorial was tested against a 500-foot grid to see if the data clustered at the block level. The z-scores showed that any apparent clustering was probably due to chance.

For point data where there is no grid available, it is possible to aggregate the data into another point feature class. This might be done using the Integrate command, grouping several of the points into a single location, keeping a count of how many features were integrated. The result is a point with summary data that can be analyzed with the Spatial Autocorrelation tool.

STUDY QUESTIONS

1. What, if anything, might be changed to find significant clustering in the data?
2. What two factors influence the results of Moran's I?
3. Why did the other pattern analysis tools find clustering in this data, while the Moran's I did not?

Another real-world example

The Moran's I may be used to examine gas-well production levels. Both the density and output of the wells are important in the calculation. Wells that cluster with high production might suggest that more wells in that same area would also be productive.

geographic information systems geographic information systems geographic infor
nic information systems geographic information systems geographic infor
nic information systems geographic information systems geographic infor
nic information systems geographic information systems geographic infor
nic informat   geographic information systems geographic infor
nic inform   eographic information systems geographic infor
nic infor   geographic information systems geographic infor
                geographic information systems geographic infor
                eographic information systems geographic infor
                geographic information systems geographic infor
                graphic infor
                on systems geographic infor
nic info   ems geographic infor
nic info   geographic infor
nic info   phic information systems geographic infor
nic inform   eographic information systems geographic infor
nic informati   geographic information systems geographic infor
nic information systems geographic information systems geographic infor
nic information systems geographic information systems geographic infor

9

Identifying clusters

Once the statistical significance of patterns in a dataset is established, it is then possible to pinpoint the locations of clustering patterns. The two methods discussed in this chapter will apply statistical values identified with the pattern analysis tools and produce a visual output that can then be used to graphically display the results. The results may be used to analyze the data's historical significance or to help predict future trends.

Tutorial 9-1

Performing cluster and outlier analysis

Checking z-scores with pattern analysis tools provides the basis for determining the probability that spatial clustering is not due to random chance. Cluster and outlier analysis is then used to display the results graphically. The Cluster and Outlier Analysis tool (Anselin local Moran's I) stores the results of the pattern analysis in the layer's attribute table, allowing for their use in symbol classification.

Learning objectives

- *Analyze spatial clustering*
- *Evaluate z-scores*
- *Identify hot spots and cold spots*
- *Classify symbols*

Preparation

- *Read pages 147–174 in* The ESRI Guide to GIS Analysis, Volume 2.

Introduction

Pattern analysis lets you determine if clustering was occurring to a high enough probability to reject the null hypothesis. Once it is determined that data is clustering, the causes of the clustering can be examined by displaying the results graphically.

The Cluster and Outlier Analysis tool will take a set of features with a weight or attribute value, perform pattern analysis, and display the results in such a way as to highlight the clustering. Areas where high values cluster are called hot spots, and areas where low values cluster are called cold spots. When these are displayed graphically, they can be overlaid on other datasets to help identify contributing factors.

The Cluster and Outlier tool calculates several new values that are used to analyze the data, which are added to new fields in the output feature class. The LMiIndex indicates whether the feature has other features of similar values near it. Positive numbers represent features that are part of a cluster, and negative values represent features that are spatial outliers. The LMiZScore will show the statistical significance of the clustering, as described earlier, and the LMiPValue is the probability of error in rejecting the null hypothesis. The COType field distinguishes between a statistically significant (0.05 level) cluster of high values (HH), cluster of low values (LL), outlier high value surrounded primarily by low values (HL), and outlier low value surrounded primarily by high values (LH).

Scenario   After running several spatial pattern analyses, the fire chief needs a map showing the results for the next city council meeting. The chief would also like to show hot and cold spots based on feature locations and the ranking of the calls. Then the data can be overlaid on the census data to see if there is some type of relationship between the two datasets.

Data   The first dataset is Calls for Service data that the fire department has compiled for February 2007. In addition, there is an attribute field that contains a number from 1 to 10 to designate the priority of the call.

The second set of data is the Census 2000 block-group-level data. It contains a field for total population, which can be used in a quantity classification. The area we need is cut out for us to use, but the analysis could be repeated for any area with the data from the media kit.

The third set of data is the street network data to give the map some reference. It comes from the ESRI Data & Maps media.

Run the Cluster and Outlier Analysis tool

1   **In ArcMap, open Tutorial 9-1.mxd.** Here's the familiar Battalion 2 data. First you will run the Local Moran's I Analysis tool to show where the calls for service might be clustered, then display the census data with it to see if there is a relationship between the two.

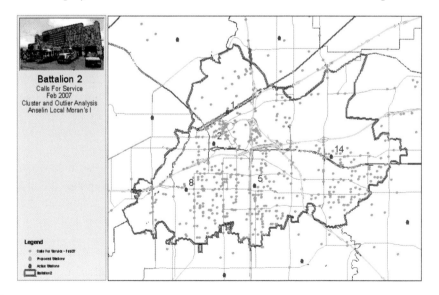

2   Use the Search window to locate and open the Cluster/Outlier Analysis with Rendering tool.

9-1

9-2

3 Enter parameters as follows:

- **Input Feature Class: Calls for Service–Feb07**
- **Input Field: FEE**
- **Output Layer File: \GIST2\MyExercises\MoransICluster**
- **Output Feature Class: \GIST2\MyExercises\MyData.mdb\MoransICluster**

4 When your dialog box matches the image, click OK. When the tool finishes, it will create a new feature class with the results of the Moran's I analysis. This includes values for the LMiIndex, LMiZScore, and LmiPValue, as discussed in the introduction.

5 Open the attribute table of the Moran's I layer and note the new values that were added. When you have finished, close the table.

In the newly symbolized feature class, the red dots represent hot spots, or areas where features have high positive z-scores because they are surrounded by features with similar values, high or low. The blue dots represent the cold spots, or areas where features have low negative z-scores because they are surrounded by features with dissimilar values.

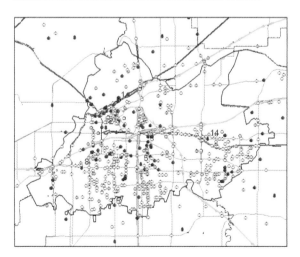

Since high z-scores can represent both high and low values of the input field, it would help to know where high values are surrounded by other high values, or low values are surrounded by other low values. An additional field is added to the results table to show where this occurs.

6 Create a copy of the layer MoransICluster in the table of contents and rename it **ClusterTypes**. Set the symbology to Unique Values and the Value Field to COType.

7 Add the values HH, HL, LH, and LL. Uncheck all other values.

8 Change the symbols to Circle 1 with a size of 50. Change the color of HH to Fire Red, LL to Leaf Green, and both HL and LH to Solar Yellow.

9 Set the transparency of the layer to 40%. Close Layer Properties.

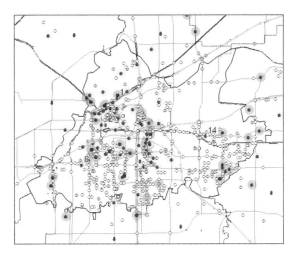

10 Move the ClusterTypes layer below the MoransICluster layer. The red symbols will show where high values are clustering with other high values. The green symbols show where low values are clustering with other low values. The yellow symbols show where high and low values are clustering. Areas where there was not significant clustering of values do not get a COType value and therefore do not appear in this map.

The hot and cold spots are shown, but what is the cause of this phenomenon? One reason might be that the areas with the highest density of population cause the most calls for service.

11 Turn on the Census 2000 layer. Look for areas with hot or cold spots and see if you can determine a relationship with the census data. Some hot spots are occurring in areas with a high population, and some are occurring in areas with a medium population. The cold spots seem to be occurring in all the population ranges. After visually assessing the data, it seems that there is no correlation between these two datasets. Further investigation with other data layers might give better insight into any underlying causes for the clustering.

Rerun the Cluster and Outlier Analysis tool with a fixed distance band

The Moran's I tool with rendering used inverse distance for the conceptualization of spatial relationships. This is not always the best choice because you are not able to control the search distance it uses to find a significant number of neighbors. In the previous tutorials, you plotted the z-score for various distances to determine which of them produce the best results in regard to nearby values. You will run the Moran's I tool again without the rendering and set a fixed distance band according to what you discovered before with the pattern analysis tools.

1 Turn off the Census 2000, Cluster Types, and MoransICluster layers.

2 Locate and open the Cluster and Outlier Analysis: Anselin Local Moran's I tool (use the tool without rendering).

3 Enter parameters as follows:

- Input Feature Class: Calls for Service–Feb07
- Input Field: FEE
- Output Feature Class: \GIST2\MyExercises\MyData.mdb\MoransICluster900
- Conceptualization of Spatial Relationships: Fixed Distance Band
- Distance Band: 900 feet

4 When your dialog box matches the image, click OK. The resulting feature class is added to the table of contents, and several new fields have been added to the attribute table. The field LMiIndex has the resulting Moran's I index value. This can be symbolized the same as the first layer you created to show hot spots and cold spots. Remember that positive values show clustering, and negative numbers show dispersion.

Cluster and Outlier Analysis (Anselin Local Morans I)

Input Feature Class
Calls For Service - Feb07

Input Field
FEE

Output Feature Class
C:\ESRIPress\GIST2\MyExercises\MyData.mdb\MoransICluster900

Conceptualization of Spatial Relationships
Fixed Distance Band

Distance Method
Euclidean Distance

Standardization
None

Distance Band or Threshold Distance (optional)
900

Weights Matrix File (optional)

OK   Cancel   Environments...   Show Help >>

5 Open the properties of the MoransICluster900 layer and import the symbology from the MoransICluster_Layer file. Use the field LmiZScore Fixed 900 for the value field. Close the layer properties and turn on the Alarm Territories layer. Compare the new results with the results of the first Cluster and Outlier analysis. The hot spots seem to correspond to portions of the major freeways, with a particular intersection in District 5 showing up as very dangerous. Perhaps freeways are one of the underlying causes of the hot spots.

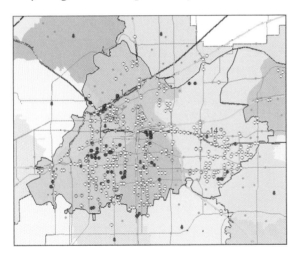

6 To determine which of the high z-scores were created by high values being close to other high values and which z-scores were the results of low values being close to other low values, you should set up a copy of layer MoransICluster900 and symbolize the COType field as before:

- Copy the MoransICluster900 layer.
- Name the copied layer ClusterTypes900.
- Change symbology type to Unique Values.
- Set the Value Field to COType.
- Add the values HH, HL, LH, and LL, and uncheck all other values.
- Change the symbol to Circle 1 with a size of 50.
- Change circle colors.
- Set transparency to 40%.
- Move the layer below MoransICluster900 in the table of contents.

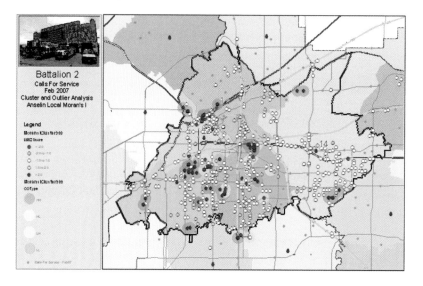

7 Save your map document as **Tutorial 9-1.mxd** in the \GIST2\MyExercises folder. If you are not continuing on to the exercise, exit ArcMap.

Exercise 9-1

The tutorial showed how to create a feature class to demonstrate the results of the Cluster and Outlier Analysis tool. From this, the data can be compared with other datasets to search for a correlation.

In this exercise, you will repeat the process using different datasets. The Incident_Jan 07 data can also be used for this type of analysis, using the FEE values for the weight.

- Open Tutorial 9-1E.mxd.
- Run the Cluster and Outlier Analysis tool. Use FEE values as the weight.
- Increase the symbol size of the output, and symbolize the CensusBlkGrpIncome layer with graduated colors using the P053001 field (determined by the census metadata) to represent Median Household Income.
- Create symbology showing cluster types.
- Add a text box discussing any possible relationships between the calls for service and median household income.
- Change the titles, colors, and legend to make a visually pleasing map.
- Repeat the analysis using the Cluster and Outlier tool with a fixed distance band of 900 feet to reduce the threshold errors.
- Create the symbology showing the cluster types for this new layer as well.
- Save the results as **Exercise 9-1.mxd** in the \GIST2\MyExercises folder.

WHAT TO TURN IN

If you are working in a classroom setting with an instructor, you may be required to submit the maps you created in tutorial 9-1.

Turn in a printed map or screen capture of the following:

Tutorial 9-1.mxd (2 maps)

Exercise 9-1.mxd (2 maps)

Tutorial 9-1 review

The Cluster and Outlier Analysis tool ran the pattern-analysis calculations and symbolized the results. The Moran's I index was calculated to show the level of clustering, with the positive values showing clustering and the negative values showing dispersion. The z-scores were also calculated for the features to help you decide if the clustering was statistically significant. High z-scores may result from features with high values being surrounded by other features with high values, or from features with low values surrounded by other features with low values. Features with low values surrounded by features with high values, or conversely features with high values surrounded by features with low values, produce a low z-score. The field COType was used to determine the type of clustering observed.

The results were also displayed over another dataset to try to visually determine if any underlying factors existed. The Census 2000 data showed total population and was compared to the hot spots. No association seemed to exist, so perhaps other factors should be investigated. The exercise included income levels and provided you another possible factor in the clustering.

STUDY QUESTIONS

1. What relationship does the z-score have to the Moran's I index?

2. What other data might be having an effect on the calls for service?

3. One analysis used the inverse distance as the conceptualization of spatial relationships, and the other used a fixed distance band. What difference did this make?

Other real-world examples

An oil-well drilling company may look at the clustering of high output values to site new wells. Areas with clustering of low values might be avoided.

A testing agency might look for clustering of high school test scores to find schools that excel, or schools that do poorly, to see if there is an association between the area and the scores.

Tutorial 9-2

Performing hot-spot analysis

The Getis-Ord hot-spot analysis shows where high and low values are clustered. The tool compares the values of each feature with the neighboring features within a user-specified distance. The values for each feature are then color-coded to show high- and low-value clusters.

Learning objectives

- *Evaluate z-scores*
- *Analyze high- and low-value clustering*
- *Classify symbols*

Preparation

- *Read pages 175–190 in* The ESRI Guide to GIS Analysis, Volume 2.

Introduction

The Gi statistic, also called Gi* (pronounced G-I-star), uses both the location and the value in the pattern calculations. This is used to see the effect of the value field on the clustering over the user specified distance. The distance is determined by the characteristics of the input dataset. Features representing large, wide groupings may use a larger distance band value, while features representing a local region or smaller feature areas might use a small distance band value. With a larger value, expect to get a few large clusters. A smaller distance value may result in many, smaller clusters.

Once the calculations are completed, the results are symbolized to display the clustering of high and low values. The Hot Spot Analysis tool sets the classification for the new layer with the low-value clusters shown in dark blue and the high-value clusters shown in red.

Scenario   The Tarrant County Economic Development Office has gotten word of the great maps you made for the fire department and wants to get in on the act. It would like to see where the median income per household is clustering. It would like to direct charity-collection efforts to areas with the largest clusters of high income, and target new job creation in areas with the lowest median income. The charity efforts will help support many county initiatives, and the job opportunities may help raise the standard of living of struggling families.

You will need to create a map showing the areas where the median income values cluster using the Gi* statistic.

Data   The first set of data is the Census 2000 block-group-level data. It contains a field for the median income, P053001, which can be used in the Gi* function. The area we need is cut out for us to use, but the analysis could be repeated for any area with the data from the media kit.

The second set of data is the street network data to give the map some reference. It also comes from the ESRI Data & Maps media.

Map high-income and low-income clusters

1   **In ArcMap, open Tutorial 9-2.mxd.** The map shows a very simple set-up. The county roads are shown with the Census 2000 data, which includes income. The Gi* function can be run on the Census2000IncomeTarr layer with the P053001 field as the input value.

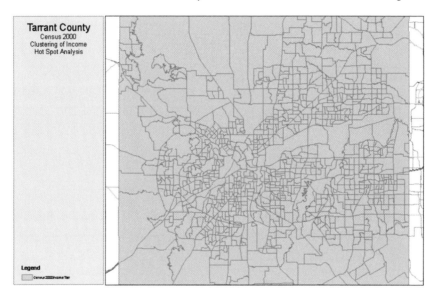

2   Use the Search window to locate and open the Hot Spot with Rendering tool.

3   Enter parameters as follows:

- **Input Feature Class: Census2000IncomeTarr**
- **Input Field: P053001 (median income)**
- **Output Layer File: \GIST2\MyExercises\GetisTarrant**
- **Output Feature Class: \GIST2\MyExercises\MyData.mdb\GetisTarrant**
- **Distance Band: 5280 feet (1 mile)**

4 When your dialog box matches the image, click OK. A new feature class is created and added to the table of contents. This will automatically symbolize the results of the Gi* statistic.

5 Make sure the Census2000IncomeTarr layer is turned off and the GetisTarrant layer is visible. The dark blue areas represent clustering of low values, and the red areas represent clustering of high values. This is a pretty dramatic result, giving a clear indication where to concentrate the job-creation programs and the charity-collection efforts.

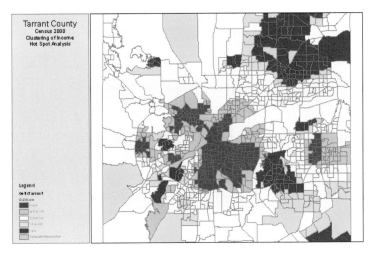

6 Save your map document as **Tutorial 9-2.mxd** in the \GIST2\MyExercises folder. If you are not continuing on to the exercise, exit ArcMap.

9-1

9-2

Exercise 9-2

The tutorial showed how to create a feature class to demonstrate the results of the Gi* Analysis tool. The results were automatically symbolized to show high- and low-value clustering.

In this exercise, you will repeat the process using different datasets. The Dallas County data is provided for doing the same Gi* analysis, using the field P053001 for the value field.

- Open Tutorial 9-2E.mxd.
- Run the Gi* Hot Spot Analysis tool. Use the P053001 field as the value.
- Set the Distance Band Value to 5280 feet.
- Add a text box with a note describing the significance of the results.
- Change the titles, colors, and legend to make a visually pleasing map.
- Save the results as **Exercise 9-2.mxd** in the \GIST2\MyExercises folder.

WHAT TO TURN IN

If you are working in a classroom setting with an instructor, you may be required to submit the maps you created in tutorial 9-2.

Turn in a printed map or screen capture of the following:

Tutorial 9-2.mxd

Exercise 9-2.mxd

Tutorial 9-2 review

The Gi* Hot Spot Analysis tool was very simple in its execution and very dramatic in the results. The median household income was used as the input value, along with the distance band value, to determine how one feature compared to its neighbors. Areas where high values clustered were symbolized in red, and areas where low values clustered were shown in blue.

This clarity of the results was due to the choice of an optimal distance band value. The high value resulted in a few large areas of clustering. This will help the county plan its efforts more effectively.

STUDY QUESTIONS

1   What results might you expect with a smaller distance band value?

2   How does Gi* differ from the Moran's I Clustering tool?

Other real-world examples

Police departments use Gi* to identify hot spots of crime by severity or frequency. The crimes can be ranked with the highest numbers assigned to felonies and the lowest assigned to misdemeanors. The Getis-Ord Gi* analysis will show where the higher ranks cluster and where the lower ranks cluster. Crimes may also be aggregated to regions such as city blocks and analyzed with the Getis-Ord Gi* tool to show the hot or cold spots of crime frequency.

The city housing authority may geocode the addresses of reported foreclosures within the city limit. These locations can be aggregated by census blocks, then analyzed with the Getis-Ord Gi* tool. The hot spots for foreclosures can be overlaid on various demographic information to investigate underlying causes.

Independent projects

The previous nine chapters with seventy tutorials and exercises have demonstrated many analysis techniques. These have ranged from simple visual analysis to complex spatial statistics. Now is your turn to demonstrate mastery of these concepts.

At the beginning of each chapter, there is a list of learning objectives. These are things that everyone who completes the book should be able to perform.

This final chapter can be used as a class project or for individuals to challenge themselves. It involves taking the data provided on the DVD with this book and developing an analysis project from scratch. The steps should include the following:

- Presenting the idea
- Outlining the process
- Validating the data
- Performing the analysis
- Validating the outcome
- Presenting the results

Choose one of the scenarios given below or write one of your own. Then perform all the steps for the analysis project listed above. All of the scenarios use Fort Worth Fire Department data for Battalion 2.

Scenario 1

As a quick visual analysis, the fire chief would like to see each month's call-for-service data categorized by call severity. Use the FEE field, with the highest number being the most severe. Overlay a 1-mile ring around all the stations in Battalion 2 and summarize the calls by severity for each station. Include a table showing the results on the map. Make a map series or animation of March 2006 to February 2007.

Scenario 2

The fire chief utters one simple question that is actually a pretty complex analysis project: "How many calls did we run in Battalion 2 that were out of a station's regular service area?"

In the incident data, the field "station" contains the number of the station that responded. Using the polygon boundary of service areas, you should be able to answer this question. Make a map series or animation of March 2006 to February 2007.

Scenario 3

The fire chief again tosses out another complex question: "What's the average drive time to a call for each station?"

Use the Tarrant Roads centerline file to make a network database. Then use the Network Analyst tools to map the route from each station to every call it made, and get the average time in minutes for each route assuming an average driving speed of 40 miles per hour. Make a map series or animation of March 2006 to February 2007.

Scenario 4

As a review for the city's insurance group, you need to make a map that shows drive time along a network from each of the stations in Battalion 2. The times should be one, two, and three minutes using an average speed of 40 miles per hour. Then you need to summarize how many responses were within each of these distances, and how many were outside these distances. These can be totals for Battalion 2, including all its stations—for instance, all the calls within one minute of a station, all calls within two minutes of a station, all the calls within three minutes of a station, and all the calls outside three minutes of a station. Make a map series or animation of March 2006 to February 2007.

Scenario 5

Demonstrate how the mean weighted center for calls for service of each station moves throughout the year. Show the movement of the weighted standard deviational ellipse for each station as well.

Make a map series or animation of March 2006 to February 2007.

Scenario 6

Determine a suitable z-score and perform a hot-spot analysis for each month's incident data. Make a map series or animation of March 2006 to February 2007.

Reference materials

- *Class notes*
- *ArcGIS Online Help*
- *The ESRI Guide to GIS Analysis, Volumes 1 and 2*
- *Tutorials and exercises*

Appendix A

Task index

Software tool/concept, **tutorial(s) in which it appears**

Appendix B

Data source credits

Chapter 1 data sources include

\ESRIPress\GIST2\Data\City Of Oleander.mdb\Land Records\LotBoundaries, derived
from City of Euless.

\ESRIPress\GIST2\Data\City Of Oleander.mdb\Land Records\Zoning, derived from City
of Euless.

\ESRIPress\GIST2\Data\City Of Oleander.mdb\Land Records\Parcels, derived from City
of Euless.

Chapter 2 data sources include

\ESRIPress\GIST2\Data\Census.mdb\MajorRoads, from ESRI Data & Maps, 2005,
courtesy of TANA/GDT.

\ESRIPress\GIST2\Data\Census.mdb\DFWRegion\CensusBlkGrp, from ESRI Data &
Maps, 2005, courtesy of TANA/GDT, US Census, ESRI BIS(Pop2004).

\ESRIPress\GIST2\Data\FoodStoresHispanic.csv, created by author.

\ESRIPress\GIST2\Data\FoodStoresHispanic.xls, created by author.

\ESRIPress\GIST2\Data\City Of Oleander.mdb\Land Records\Parcels, derived from City
of Euless.

\ESRIPress\GIST2\Data\City Of Oleander.mdb\Land Records\LotBoundaries, derived
from City of Euless.

Chapter 3 data sources include

\ESRIPress\GIST2\Data\Census.mdb\MajorRoads, from ESRI Data & Maps, 2005,
courtesy of TANA/GDT.

\ESRIPress\GIST2\Data\Census.mdb\DFWRegion\CensusBlkGrp, from ESRI Data &
Maps, 2005, courtesy of TANA/GDT, US Census, ESRI BIS(Pop2004).

\ESRIPress\GIST2\Data\City Of Oleander.mdb\Land Records\City_Limit, derived from
City of Euless.

\ESRIPress\GIST2\Data\City Of Oleander.mdb\TreeInventory, derived from City of Euless.

Chapter 4 data sources include

\ESRIPress\GIST2\Data\City Of Oleander.mdb\Land Records\LotBoundaries, derived from City of Euless.

\ESRIPress\GIST2\Data\City Of Oleander.mdb\Land Records\Parcels, derived from City of Euless.

\ESRIPress\GIST2\Data\City Of Oleander.mdb\FloodPlane\FloodAreas, derived from City of Euless.

\ESRIPress\GIST2\Data\City Of Oleander.mdb\FloodPlane\FloodZone, derived from City of Euless.

Chapter 5 data sources include

\ESRIPress\GIST2\Data\City Of Oleander.mdb\Land Records\LotBoundaries, derived from City of Euless.

\ESRIPress\GIST2\Data\City Of Oleander.mdb\Land Records\Parcels, derived from City of Euless.

\ESRIPress\GIST2\Data\Census.mdb\TarrantCounty\TarrantRivers, derived from North Central Texas Council of Governments.

\ESRIPress\GIST2\Data\City Of Oleander.mdb\DFWCities, derived from North Central Texas Council of Governments.

\ESRIPress\GIST2\Data\City Of Oleander.mdb\Land Records\Lot_Buffer, created by author.

\ESRIPress\GIST2\Data\Networks.mdb\NetworkAnalysis\StreetCenterlines, derived from City of Euless.

\ESRIPress\GIST2\Data\City Of Oleander.mdb\FireDepartment\Stations, derived from City of Euless.

\ESRIPress\GIST2\Data\City Of Oleander.mdb\Planimetric Data\BuildingFootprints, derived from City of Euless.

\ESRIPress\GIST2\Data\Census.mdb\MajorRoads, from ESRI Data & Maps, 2005, courtesy of TANA/GDT.

\ESRIPress\GIST2\Data\City Of Oleander.mdb\FireDepartment\FireRuns0505, derived from City of Euless.

Chapter 6 data sources include

\ESRIPress\GIST2\Data\City Of Oleander.mdb\Land Records\LotBoundaries, derived from City of Euless.

\ESRIPress\GIST2\Data\City Of Oleander.mdb\Planimetric Data\BuildingFootprints, derived from City of Euless.

\ESRIPress\GIST2\Data\City Of Oleander.mdb\FireDepartment\Level1, created by author.

\ESRIPress\GIST2\Data\City Of Oleander.mdb\FireDepartment\Level2, created by author.

\ESRIPress\GIST2\Data\City Of Oleander.mdb\FireDepartment\Level3, created by author.

\ESRIPress\GIST2\Data\City Of Oleander.mdb\FireDepartment\Site2_Level1, created by author.

\ESRIPress\GIST2\Data\City Of Oleander.mdb\FireDepartment\Site2_Level2, created by author.

\ESRIPress\GIST2\Data\Census.mdb\MajorRoads, from ESRI Data & Maps, 2005, courtesy of TANA/GDT.

\ESRIPress\GIST2\Data\City Of Oleander.mdb\FireDepartment\TornadoApril07, created by author.

\ESRIPress\GIST2\Data\City Of Oleander.mdb\FireDepartment\TornadoDrill, created by author.

\ESRIPress\GIST2\Data\City Of Oleander.mdb\FireDepartment\TornadoPath_Buffer, created by author.

\ESRIPress\GIST2\Data\City Of Oleander.mdb\FireDepartment\TornadoPathApril07, created by author.

\ESRIPress\GIST2\Data\City Of Oleander.mdb\FireDepartment\TornadoPathDrill, created by author.

\ESRIPress\GIST2\Data\City Of Oleander.mdb\Land Records\ParcelTaxValue, derived from City of Euless.

Chapter 7 data sources include

\ESRIPress\GIST2\Data\Census.mdb\MajorRoads, from ESRI Data & Maps, 2005, courtesy of TANA/GDT.

\ESRIPress\GIST2\Data\City of Ft Worth.mdb\Fire Department\Active_Stations, derived from City of Fort Worth.

\ESRIPress\GIST2\Data\City of Ft Worth.mdb\Fire Department\Alarm_Territories, derived from City of Fort Worth.

\ESRIPress\GIST2\Data\City of Ft Worth.mdb\Fire Department\Battalions, derived from City of Fort Worth.

\ESRIPress\GIST2\Data\City of Ft Worth.mdb\Fire Department\Incident_Feb07, derived from City of Fort Worth.

\ESRIPress\GIST2\Data\City of Ft Worth.mdb\Fire Department\Proposed_Stations, derived from City of Fort Worth.

\ESRIPress\GIST2\Data\Library.mdb\Districts, created by author.

\ESRIPress\GIST2\Data\Library.mdb\PatronLocations, derived from City of Euless.

\ESRIPress\GIST2\Data\City Of Oleander.mdb\Land Records\LotBoundaries, derived from City of Euless.

\ESRIPress\GIST2\Data\City Of Oleander.mdb\Land Records\Parcels, derived from City of Euless.

\ESRIPress\GIST2\Data\Census.mdb\DFWRegion\CensusBlkGrp, from ESRI Data & Maps, 2005, courtesy of TANA/GDT, US Census, ESRI BIS(Pop2004).

\ESRIPress\GIST2\Data\City of Ft Worth.mdb\Fire Department\Storm_Track, derived from city of Fort Worth.

\ESRIPress\GIST2\Data\City of Ft Worth.mdb\Fire Department\Tornado, derived from City of Fort Worth.

\ESRIPress\GIST2\Data\City of Ft Worth.mdb\Fire Department\Storm_Track_Dallas, derived from City of Fort Worth.

\ESRIPress\GIST2\Data\City of Ft Worth.mdb\Fire Department\Tornado_Dallas, derived from City of Fort Worth.

Chapter 8 data sources include

\ESRIPress\GIST2\Data\Census.mdb\MajorRoads, from ESRI Data & Maps, 2005, courtesy of TANA/GDT.

\ESRIPress\GIST2\Data\City of Ft Worth.mdb\Fire Department\Active_Stations, derived from City of Fort Worth.

\ESRIPress\GIST2\Data\City of Ft Worth.mdb\Fire Department\Alarm_Territories, derived from City of Fort Worth.

\ESRIPress\GIST2\Data\City of Ft Worth.mdb\Fire Department\Battalions, derived from City of Fort Worth.

\ESRIPress\GIST2\Data\City of Ft Worth.mdb\Fire Department\Incident_Feb07, derived from City of Fort Worth.

\ESRIPress\GIST2\Data\City of Ft Worth.mdb\Fire Department\Proposed_Stations, derived from City of Fort Worth.

\ESRIPress\GIST2\Data\City of Ft Worth.mdb\Fire Department\Battalion2, derived from City of Fort Worth.

\ESRIPress\GIST2\Data\City of Ft Worth.mdb\Fire Department\Incident_Jan07, derived from City of Fort Worth.

\ESRIPress\GIST2\Data\City of Ft Worth.mdb\Fire Department\Grid500, derived from City of Fort Worth.

\ESRIPress\GIST2\Data\City of Ft Worth.mdb\Fire Department\Grid200, derived from City of Fort Worth.

Chapter 9 data sources include

\ESRIPress\GIST2\Data\Census.mdb\MajorRoads, from ESRI Data & Maps, 2005, courtesy of TANA/GDT.

\ESRIPress\GIST2\Data\Census.mdb\TarrantCounty\Census2000Ethnic, derived from North Central Texas Council of Governments.

\ESRIPress\GIST2\Data\City of Ft Worth.mdb\Fire Department\Active_Stations, derived from City of Fort Worth.

\ESRIPress\GIST2\Data\City of Ft Worth.mdb\Fire Department\Alarm_Territories, derived from City of Fort Worth.

\ESRIPress\GIST2\Data\City of Ft Worth.mdb\Fire Department\Battalions, derived from City of
  Fort Worth.

\ESRIPress\GIST2\Data\City of Ft Worth.mdb\Fire Department\Incident_Feb07, derived from City of
  Fort Worth.

\ESRIPress\GIST2\Data\City of Ft Worth.mdb\Fire Department\Proposed_Stations, derived from City
  of Fort Worth.

\ESRIPress\GIST2\Data\City of Ft Worth.mdb\Fire Department\Battalion2, derived from City of
  Fort Worth.

\ESRIPress\GIST2\Data\City of Ft Worth.mdb\Fire Department\Incident_Jan07, derived from City of
  Fort Worth.

\ESRIPress\GIST2\Data\City of Ft Worth.mdb\Fire Department\Census2000IncomeTarr, ESRI Data &
  Maps, 2005, courtesy of TANA/GDT, US Census, ESRI BIS(Pop2004).

\ESRIPress\GIST2\Data\City of Ft Worth.mdb\Fire Department\Census2000IncomeDall, ESRI Data &
  Maps, 2005, courtesy of TANA/GDT, US Census, ESRI BIS(Pop2004).

Independent projects data sources include
\ESRIPress\GIST2\Data\City of Ft Worth.mdb\Fire Department\Incident_Apr06, derived from City of
  Fort Worth.

\ESRIPress\GIST2\Data\City of Ft Worth.mdb\Fire Department\Incident_Aug06, derived from City of
  Fort Worth.

\ESRIPress\GIST2\Data\City of Ft Worth.mdb\Fire Department\Incident_Dec06, derived from City of
  Fort Worth.

\ESRIPress\GIST2\Data\City of Ft Worth.mdb\Fire Department\Incident_Feb07, derived from City of
  Fort Worth.

\ESRIPress\GIST2\Data\City of Ft Worth.mdb\Fire Department\Incident_Jan07, derived from City of
  Fort Worth.

\ESRIPress\GIST2\Data\City of Ft Worth.mdb\Fire Department\Incident_Jul06, derived from City of
  Fort Worth.

\ESRIPress\GIST2\Data\City of Ft Worth.mdb\Fire Department\Incident_Jun06, derived from City of
  Fort Worth.

\ESRIPress\GIST2\Data\City of Ft Worth.mdb\Fire Department\Incident_Mar06, derived from City of
  Fort Worth.

\ESRIPress\GIST2\Data\City of Ft Worth.mdb\Fire Department\Incident_May06, derived from City of
  Fort Worth.

\ESRIPress\GIST2\Data\City of Ft Worth.mdb\Fire Department\Incident_Nov06, derived from City of
  Fort Worth.

\ESRIPress\GIST2\Data\City of Ft Worth.mdb\Fire Department\Incident_Oct06, derived from City of
  Fort Worth.

\ESRIPress\GIST2\Data\City of Ft Worth.mdb\Fire Department\Incident_Sep06, derived from City of
  Fort Worth.

Appendix C

Data license agreement

Important:
Read carefully before opening the sealed media package

Environmental Systems Research Institute Inc. (ESRI) is willing to license the enclosed data and related materials to you only upon the condition that you accept all of the terms and conditions contained in this license agreement. Please read the terms and conditions carefully before opening the sealed media package. By opening the sealed media package, you are indicating your acceptance of the ESRI License Agreement. If you do not agree to the terms and conditions as stated, then ESRI is unwilling to license the data and related materials to you. In such event, you should return the media package with the seal unbroken and all other components to ESRI.

ESRI license agreement

This is a license agreement, and not an agreement for sale, between you (Licensee) and Environmental Systems Research Institute Inc. (ESRI). This ESRI License Agreement (Agreement) gives Licensee certain limited rights to use the data and related materials (Data and Related Materials). All rights not specifically granted in this Agreement are reserved to ESRI and its Licensors.

Reservation of Ownership and Grant of License: ESRI and its Licensors retain exclusive rights, title, and ownership to the copy of the Data and Related Materials licensed under this Agreement and, hereby, grant to Licensee a personal, nonexclusive, nontransferable, royalty-free, world-wide license to use the Data and Related Materials based on the terms and conditions of this Agreement. Licensee agrees to use reasonable effort to protect the Data and Related Materials from unauthorized use, reproduction, distribution, or publication.

Proprietary Rights and Copyright: Licensee acknowledges that the Data and Related Materials are proprietary and confidential property of ESRI and its Licensors and are pro-tected by United States copyright laws and applicable international copyright treaties and/or conventions.

Permitted Uses: Licensee may install the Data and Related Materials onto permanent storage device(s) for Licensee's own internal use.

Licensee may make only one (1) copy of the original Data and Related Materials for archival purposes during the term of this Agreement unless the right to make additional copies is granted to Licensee in writing by ESRI.

Licensee may internally use the Data and Related Materials provided by ESRI for the stated purpose of GIS training and education.

Uses Not Permitted: Licensee shall not sell, rent, lease, sublicense, lend, assign, time-share, or transfer, in whole or in part, or provide unlicensed Third Parties access to the Data and Related Materials or portions of the Data and Related Materials, any updates, or Licensee's rights under this Agreement.

Licensee shall not remove or obscure any copyright or trademark notices of ESRI or its Licensors.

Term and Termination: The license granted to Licensee by this Agreement shall commence upon the acceptance of this Agreement and shall continue until such time that Licensee elects in writing to discontinue use of the Data or Related Materials and terminates this Agreement. The Agreement shall automatically terminate without notice if Licensee fails to comply with any pro-vision of this Agreement. Licensee shall then return to ESRI the Data and Related Materials. The parties hereby agree that all provisions that operate to protect the rights of ESRI and its Licensors shall remain in force should breach occur.

Disclaimer of Warranty: The Data and Related Materials contained herein are provided "as-is," without warranty of any kind, either express or implied, including, but not limited to, the implied warranties of merchantability, fitness for a particular purpose, or noninfringement. ESRI does not warrant that the Data and Related Materials will meet Licensee's needs or expectations, that the use of the Data and Related Materials will be uninterrupted, or that all nonconformities, defects, or errors can or will be corrected. ESRI is not inviting reliance on the Data or Related Materials for commercial planning or analysis purposes, and Licensee should always check actual data.

Data Disclaimer: The Data used herein has been derived from actual spatial or tabular information. In some cases, ESRI has manipulated and applied certain assumptions, analyses, and opinions to the Data solely for educational training purposes. Assumptions, analyses, opinions applied, and actual outcomes may vary. Again, ESRI is not inviting reliance on this Data, and the Licensee should always verify actual Data and exercise their own professional judgment when interpreting any outcomes.

LIMITATION OF LIABILITY: ESRI SHALL NOT BE LIABLE FOR DIRECT, INDIRECT, SPECIAL, INCIDENTAL, OR CONSEQUENTIAL DAMAGES RELATED TO LICENSEE'S USE OF THE DATA AND RELATED MATERIALS, EVEN IF ESRI IS ADVISED OF THE POSSIBILITY OF SUCH DAMAGE.

No Implied Waivers: No failure or delay by ESRI or its Licensors in enforcing any right or remedy under this Agreement shall be construed as a waiver of any future or other exercise of such right or remedy by ESRI or its Licensors.

Order for Precedence: Any conflict between the terms of this Agreement and any FAR, DFAR, purchase order, or other terms shall be resolved in favor of the terms expressed in this Agreement, subject to the government's minimum rights unless agreed otherwise.

Export Regulation: Licensee acknowledges that this Agreement and the performance thereof are subject to compliance with any and all applicable United States laws, regulations, or orders relating to the export of data thereto. Licensee agrees to comply with all laws, regulations, and orders of the United States in regard to any export of such technical data.

Severability: If any provision(s) of this Agreement shall be held to be invalid, illegal, or unenforceable by a court or other tribunal of competent jurisdiction, the validity, legality, and enforceability of the remaining provisions shall not in any way be affected or impaired thereby.

Governing Law: This Agreement, entered into in the County of San Bernardino, shall be construed and enforced in accordance with and be governed by the laws of the United States of America and the State of California without reference to conflict of laws principles. The parties hereby consent to the personal jurisdiction of the courts of this county and waive their rights to change venue.

Entire Agreement: The parties agree that this Agreement constitutes the sole and entire agreement of the parties as to the matter set forth herein and supersedes any previous agreements, understandings, and arrangements between the parties relating hereto.

Appendix D

Installing the data and software

GIS Tutorial 2: Spatial Analysis Workbook includes a DVD containing exercise and assignment data and a DVD containing a fully functioning 180-day trial version of ArcGIS Desktop 10 software (ArcEditor license level). If you already have a licensed copy of ArcGIS Desktop 10 installed on your computer (or have access to the software through a network), do not download the software. Use your licensed software to do the exercises in this book. If you have an older version of ArcGIS installed on your computer, you must uninstall it before you can install the software that is provided with this book.

.NET Framework 3.5 SP1 must be installed on your computer before you install ArcGIS Desktop 10. Some features of ArcGIS Desktop 10 require Microsoft Internet Explorer version 8.0. If you do not have Microsoft Internet Explorer version 8.0, you must install it before installing ArcGIS Desktop 10.

Installing the maps and data

Follow the steps below to install the maps and data.

1 Put the data DVD in your computer's DVD drive. A splash screen will appear.

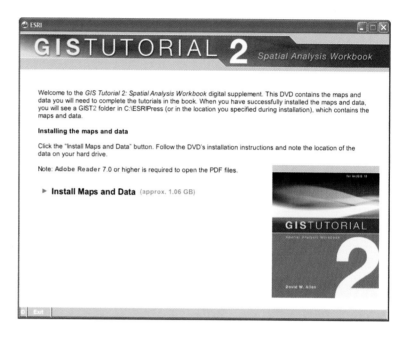

2 Read the welcome, then click the Install Maps and Data link. This launches the InstallShield Wizard.

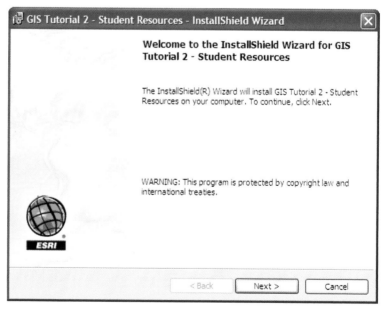

3 Click Next. Read and accept the license agreement terms, then click Next.

4 Accept the default installation folder or click Browse and navigate to the drive or folder location where you want to install the data.

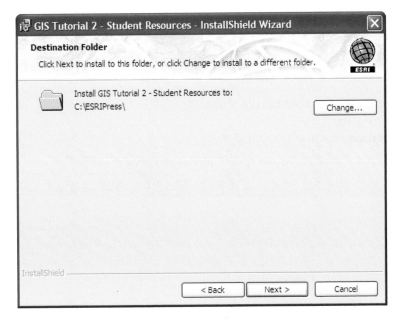

5 Click Next. The installation will take a few moments. When the installation is complete, you will see the following message.

6 Click Finish. The exercise data is installed on your computer in a folder called C:\ESRIPress\GIST2.

Uninstalling the maps and data

To uninstall the maps and data from your computer, open your operating system's control panel and double-click the Add/Remove Programs icon. In the Add/Remove Programs dialog box, select the following entry and follow the prompts to remove it:

GIS Tutorial 2

Installing the software

The ArcGIS software included with this book is intended for educational purposes only. Once installed and registered, the software will run for 180 days. The software cannot be reinstalled nor can the time limit be extended. It is recommended that you uninstall this software when it expires.

> **IMPORTANT**
>
> Before installing the trial software, visit www.esri.com/180daytrial to activate your authorization code. Visit www.esri.com/evalhelp for evaluation software support.

Follow the steps below to install the software.

1 Put the software DVD in your computer's DVD drive. A splash screen will appear. If your auto-run is disabled, navigate to the contents of the DVD and double-click the ESRI.exe file to begin.

2 Click the ArcGIS Desktop Setup installation option. This will launch the Setup wizard.

3 Read the Welcome screen, and then click Next.

4 Read the license agreement. Click "I accept the license agreement," and then click Next.

5 Choose the Complete install option, which will add extension products that are used in the book. Click Next.

6 Accept the default installation folder or navigate to the drive or folder location where you want to install the software. Click Next.

7 Accept the default installation folder or navigate to the drive or folder where you want to install Python, a scripting language used by some ArcGIS geoprocessing functions. (You won't see this panel if you already have Python installed.) Click Next.

8 The installation paths for ArcGIS and Python are confirmed. Click Next. The software will take some time to install on your computer.

9 Click Finish when the installation is completed.

10 On the ArcGIS Administrator Wizard window, select ArcEditor (Single Use), then click Authorize Now.

11 Select "I have installed my software and need to authorize it." Click Next.

12 Follow the wizard to begin the authorization process. Use the authorization code located at the bottom of the software DVD jacket in the back of the book.

If you have questions or encounter problems during the installation process, or while using this book, please use the resources listed below. (The ESRI Technical Support Department does not answer questions regarding the ArcGIS Desktop software, the GIS Tutorial supplementary media, or the contents of the book itself.)

- To resolve problems with the trial software, visit www.esri.com/evalhelp.

- To report problems with the software or exercise data DVDs, or to report mistakes in the book, send an e-mail to ESRI workbook support at workbook-support@esri.com.

Uninstalling the software

To uninstall the software from your computer, open your operating system's control panel and double-click the Add/Remove Programs icon. In the Add/Remove Programs dialog box, select the following entry and follow the prompts to remove it:

ArcGIS Desktop 10

Related titles from ESRI Press

The ESRI Guide to GIS Analysis, Volume 1: Geographic Patterns and Relationships

ISBN: 978-1-87910-206-4

This book presents the necessary tools to conduct real analysis with GIS. Using examples from various industries, this book focuses on six of the most common geographic analysis tasks: mapping where things are, mapping the most and least, mapping density, finding what is inside, finding what is nearby, and mapping what has changed.

The ESRI Guide to GIS Analysis, Volume 2: Spatial Measurements and Statistics

ISBN: 978-1-58948-116-9

As the tools available through commercial GIS software have grown in sophistication, a need has emerged to instruct users on the best practices of true GIS analysis. In this sequel to the bestselling *The ESRI Guide to GIS Analysis, Volume 1*, author Andy Mitchell delves into the more advanced realm of spatial measurements and statistics. Covering topics that range from identifying patterns and clusters, to analyzing geographic relationships, this book is a valuable resource for GIS users performing complex analysis.

GIS, Spatial Analysis, and Modeling

ISBN: 978-1-58948-130-5

Recent advancements in GIS software, along with the availability of spatially referenced data, now makes possible the sophisticated modeling and statistical analysis of all types of geographic phenomena. *GIS, Spatial Analysis, and Modeling* presents a compendium of papers that advance the methods and practices used to develop meaningful spatial analysis for decision support. A resource for geographic analysts, modelers, software engineers, and GIS professionals, this text covers tools, techniques, and methods while providing examples of various applications.

GIS Tutorial 3: Advanced Workbook

ISBN: 978-1-58948-207-4

GIS Tutorial 3 features exercises that demonstrate the advanced functionality of ArcGIS Desktop. This workbook is divided into four sections: geodatabase framework design, data creation and management, workflow optimization, and labeling and symbolizing. *GIS Tutorial 3* was designed to be used for advanced coursework or individual study.

ESRI Press publishes books about the science, application, and technology of GIS. Ask for these titles at your local bookstore or order by calling 1-800-447-9778. You can also read book descriptions, read reviews, and shop online at www.esri.com/esripress. Outside the United States, visit our Web site at www.esri.com/esripressorders for a full list of book distributors and their territories.